# Risk Management Practices in the Fire Service

This publication was produced under contract EMW-95-C-4695 for the United States Fire Administration, Federal Emergency Management Agency. Any information, findings, conclusions, or recommendations expressed in this publication do not necessarily reflect the views of the Federal Emergency Management Agency or the United States Fire Administration.

# Acknowledgment

This text is possible because of many persons who have worked and written in the field of Risk Management over the past fifteen years. The Appendix provides a list of organizations and services and a select bibliography of publications that were used in the development of this text.

Many persons have made significant contributions to this text, The following have been of particular help, for which we offer our sincere thanks: Murrey Loflin, Virginia Beach, Virginia Fire Department; Bill Jenaway, Ring of Prussia, Pennsylvania Fire Department; Anthony Granito, Emergency Service Insurance Program, Cortland, New York; Gerry Rosicky, George Rounds, Bill Solly, General Motors Corporation; Chief Jon Alberghini of the Kingston, Massachusetts Fire Department.

# Table of Contents

# About This Manual

## Purpose

---

**The manual: a comprehensive resource guide**

This manual is about **risk management.** Admittedly, the term "risk management" can be applied to a wide range of functions and activities. Several special applications of risk management, however, are uniquely significant to organizations that, in the course of delivering emergency services to the public, cope with high risk situations as an integral component of their missions. Such organizations include:.

- fire departments
- rescue squads
- emergency medical services
- other related organizations.

Risk management calls for a multi-faceted approach that encompasses many elements, some of which are:

- safety
- health
- financial impact
- loss control.

It is a fundamental tenet of risk management that it be **an ongoing, evolving, regularly refreshed and continuously improved process.**

The manual is designed as a comprehensive guide that helps fire and emergency service providers understand the concepts that form the foundation of risk management principles and practices, In addition, the manual directs the reader to sources of additional information and operational examples. The manual focuses on the practical application of risk management principles to fire department operations.

---

*Continued on next page*

## Purpose,

**How this manual is organized**

The manual is divided into the following chapters:

About This Manual
Chapter One: Overview
Chapter Two: Organizational Risk Management
Chapter Three: Operational Risk Management
Chapter Four: Managing Information
Appendix

Each chapter includes a list of topics covered in the chapter, as well as overview and summary sections.

# Overview

# Introduction

**Purpose of this Chapter**

This Chapter provides a basic risk management vocabulary and presents the operational assumptions and concepts that form the foundation of a risk management philosophy.

This Chapter helps prepare incident commanders and other emergency responders by:
- defining risk and risk management
- describing the risk management mission
- providing examples of operational risk management considerations

**In this Chapter**

This Chapter covers the following topics.

# What Is "Risk"?

**Definition**

To discuss risk management, we must first define what we mean by "risk." Unfortunately, the term "risk" has come to be used interchangeably in widely disparate settings. As a result, its meaning can be blurred in the minds of many.

A dictionary defines "risk" as a noun and a verb:
- *(noun)* The possibility of meeting danger or suffering harm or loss, or exposure to harm or loss.
- *(noun)* A person or thing insured or representing a source of risk.
- *(verb)* To expose to the chance of injury or loss.

**Determining factors**

Three interrelated determining factors are inherent in all the definitions above:
- the probability that an undesired event might occur
- a harmful or undesirable consequence
- the severity of the harm that might result.

**Describing risk**

The **probability of an occurrence** can be described in subjective terms such as "rare" and "high,' or in numerical terms such as "one in a million," "one in three," and "twenty percent."

In the same way, **harmful consequences** are often ,expressed in descriptive terms like "death," "incapacitation," "injury," "disaster," and "destruction," or in more concrete terms such as "loss of a $1.5 million facility" and "the combined costs of medical payments, compensation, pension benefits, and lost productivity" (in the event of a worker's injury).

**Probability and consequence** can be combined and expressed mathematically as the product of loss and probability, An insurance company, for example, might describe a facility as a ten million dollar risk with a two percent probability of loss. A financial analyst might calculate a ten million dollar investment that has a two percent probability of loss as a $200,000 risk.

*Continued on next page*

## What Is "Risk"?, Continued

**The probability of risk**

The probability of risk is two-pronged. It relates not only to the chance that something undesirable might happen, but also to the .probable outcome as rated on a scale of negative consequences.

For example, based on statistical calculations, we can predict the number of traffic accidents likely to occur per million miles driven. We can also predict the number of injuries and fatalities that will come about as a result. However, such a statistics-based approach will not tell us as individuals where or when we might be involved in an accident ourselves. And further, those predictive statistics certainly cannot tell us if our next accident will be a fender bender in a shopping center parking lot on a sunny afternoon or a multi-vehicle pileup on a dark two lane road on a stormy night.

**Assessing risk**

The integration of probability and consequence. helps us plan our lives and guides many of the routine personal and professional decisions we make everyday.

Out of habit, we often apply the term "risk" to virtually any type or degree of undesirable consequence or negative outcome--from a minor inconvenience to a major disaster. For instance, we might talk about the "risk" of having to share a sixty second elevator ride with an undesirable companion, or we might talk about the "risk" of an earthquake devastating a region of the United States.

Each situation carries with it a probability of occurrence and a potential degree of harm that might result. When we subjectively weigh a "risk" in terms of the undesirable consequence, we might place the outcome anywhere on a scale of undesirability. In the examples above, we might measure the negative outcome in lives lost and property damage in the millions, or simply by wishing we had waited for the next elevator.

*Continued on next page*

| | |
|---|---|
| **Relative risk** | We judge the consequences of specific unpleasant events on a broad scale of relative undesirability. For example, a heart attack can lead to hospitalization, therapy, and a long period of recovery. Or it can be fatal. Both potential outcomes are undesirable. Clearly, though, recovery is the better alternative. Often, a person who has recovered from a heart attack considers the probability and potential consequences of having another and attempts to mitigate that risk by changing exercise patterns, eating habits, and other lifestyle choices. |
| **Insurance risk** | An insurance company refers to the item, object, or person insures as "the risk." When we buy fire insurance, we make value judgments about how much we are willing to pay to be compensated for a loss if our property is damaged or destroyed by fire. When an insurance company sells us that policy, it is making a value judgment about two aspects of probability:<br>• the probability that a fire will occur<br>• the probable extent of the loss that will result. |
| **Probability, premiums, and profit** | To determine a premium, an insurance company evaluates the probability of a loss occurring and the potential size of that loss. When it establishes the rate for a fire insurance policy, a company takes into account many of the property's characteristics, including:<br>• the type of structure<br>• size<br>• value<br>• contents<br>• activities that take place there<br>• built-in fire protection systems<br>• available fire department resources.<br><br>Insurers work those factors into a prediction that considers the loss history of similar properties, the level of public fire protection and other influences, such as local climate. The rate is ultimately based on the insurer's loss experience with similar risks and the probability and the potential scope of loss for the insured property, over time. If predictions are not accurate for a specific building, the insurer should still be profitable if its predictions are generally accurate across a large number of insured risks. |

# Three Levels of Risk Management

**Introduction**    In this section, we will examine the principles and practices of risk management as they apply on three levels:
- the community as a whole
- the emergency response organization
- emergency response operations.

The term **risk management** refers to any activity that involves the evaluation or comparison of risks and the development of approaches that change the probability or the consequences of a harmful action. Risk management comprises the entire process of identification and evaluation of risks as well as the identification, selection, and implementation of **control measures** that might alter risk. Loss control is an element of risk management that is discussed later in this chapter.

Those control measures, defined in the table, can be categorized as:
- administrative
- engineering
- personnel protection.

| These Controls... | Consist of... | And are intended to... |
|---|---|---|
| Administrative | Guidelines, policies, and procedures established to limit losses. Examples include: <br>• standard operating procedures <br>• training requirements <br>• safe work practices <br>• regulations and standards | Make the task safe for the worker |
| Engineering | Engineered systems that remove or limit hazards. Examples include: <br>• apparatus design <br>• mechanical ventilation <br>• lock-out and tag-out of electrical hazards | Make the task safe for the worker |
| Personal Protection | Equipment, clothing and devices designed to protect the worker. Examples include: <br>• helmets <br>• gloves <br>• SCBA <br>• tools | Make the worker safe from the hazards |

# Risk Management in the Community

**Introduction**  By having trained personnel and specialized equipment on hand, emergency responders exemplify a community's response to risk. In fact, the basic mission of the fire, EMS, and rescue services is intimately related to the control of risk throughout a community.

**Fire control in the community**  Fire departments play a critical role in defending their communities against fires and other situations that threaten lives and property.

> Historical Note:
> Public fire departments were organized primarily to defend communities against the risk of conflagrations. Conflagrations grow from small fires that are not controlled in their early stages and grow until they cannot be controlled. Before modem concepts of fire protection were developed, conflagrations often devastated cities and towns.
>
> Over the years, our increasing ability to limit the size of fires has almost, but not entirely, eliminated the occurrence of urban conflagration. (The 1991 urban/wildland interface fire that claimed twenty-five lives and destroyed more than 2700 structures in Oakland, California, is evidence that conflagrations are still possible.)

Although most public fire departments now focus on the control of fires in individual properties and the rescue of endangered occupants, the fire suppression role of the fire department is still based on the need to protect the community's property and population. In that respect, the fire department is part of the community's fire risk management program. It exists to limit the probable loss when a fire occurs.

**Delivering fire control services to the community**  A community expresses its assessment of its overall fire risk through the resources it is willing to commit to its fire department. If the fire department is unable to perform its fire control mission, the community's fire risk balance could be compromised. The fire chief is responsible for:
- managing the community's fire risk
- providing a set of services that are part of the risk management system (the service delivery mission)
- ensuring the department can perform its mission at all times.

*Continued on next page*

# Risk Management in the Community, Continued

**Risk management-related missions**

In addition to fire control units, most of the other teams within fire departments and other emergency response organizations also conduct activities directly related to community risk management.

Emergency medical services respond to urgent situations that are related to the health and welfare of the community's citizens. Emergency management services protect the community from the effects of natural disasters. Rescue teams safely remove citizens from dangerous predicaments, avoiding the risk of injury or death that untrained, unprepared citizens might face if they tried to perform that mission. Hazardous materials response teams protect the population and the environment from the effects of uncontrolled releases of hazardous materials, The common thread among the missions of all those teams is the community's need for protection from potentially harmful or undesirable events.

Such teams play significant roles in reducing a community's vulnerability to one or more types of harm. Just the same, emergency response is typically only one dimension of their total mission.

**Risk management-related activities**

Key activities that support risk management principles and practices include fire prevention, code enforcement, public education, and other efforts that make citizens more aware of how to:
- ı recognize potentially dangerous situations
- ı prevent emergencies
- ı respond effectively to an emergency.

Communities benefit from educational and motivational programs such as:
- hazard recognition, avoidance, and elimination
- standards that require automatic sprinkler systems and fire-resistive construction
- inspection and compliance enforcement.

Taken together, all these activities reduce a community's vulnerability and are an integral part of a community's fire risk management system.

# Organizational Risk Management

| | |
|---|---|
| **Introduction** | The responsibility of a manager to safeguard the assets of his or her organization are just as applicable to an emergency response agency as to any business or family. Although their mission is to manage community risk, fire departments and other emergency service organizations must also be concerned with risk to themselves. In several different ways, those risks can keep the organization from successfully carrying out its mission. The fire department is open to a variety of risks similar to those faced by every private organization and enterprise.

There are interesting parallels between the two sectors. A risk manager in a <u>private</u> enterprise tries to protect the assets of the enterprise and ensure that it can stay in business. Similarly, a risk manager in a fire department tries to protect <u>public</u> assets, including its personnel, facilities, and equipment--and make sure the fire department can perform its mission. |
| **Public and private risks--a comparison** | By the same token, the nature of the risks each organization faces and the impact of loss can be quite different. Just as a fire in a restaurant can put it out of business, a fire at a fire station can compromise the capacity of that fire department to respond to fires. Many of the assets of the fire department are just as essential to the ability to deliver emergency services as a kitchen is to a restaurant's ability to serve food. The critical difference--if a restaurant loses business, its employees might be out of work, but the patrons can find someplace else to eat. If a fire station burns down, the community as a whole may be without adequate fire protection or emergency medical service. |
| **Emergency response as an essential public service** | Emergency response organizations are viewed as essential public services. Because that is so, emergency response managers should recognize that they are responsible for ensuring that their organizations are always ready and able to perform their missions, More than the average citizen or business manager, emergency response managers must be aware of potential disruptions to service. After all, they are expected to recognize and manage risk as part of their everyday professional lives. |

*Continued on next page*

# Organizational Risk Management, Continued

**A public trust**    The managers of public safety organizations are also custodians of public funds and assets. They must restrict any undesirable outcome that costs money, consumes public dollars, and reduces government's capability to spend those funds where they would do the most good.

For example, an agency could run up large bills replacing or repairing damaged, lost, or stolen equipment and apparatus, It could incur high expenses paying liability claims for vehicle accidents, paying medical bills for injured members, legal expenses to defend against claims, and paying overtime replacing them. In other words--the agency could be spending money to compensate for circumstances that should have been avoided.

It is the manager's responsibility to prevent such things from happening.

# Operational Risk Management

**Introduction**

This manual includes a section that deals specifically with the application of risk management concepts to the operational practices of fire departments and emergency response organizations. In that section, we address issues involved in incorporating a "Risk Management Approach" within an organization's emergency activities The approach that is outlined--one that limits or reduces the risk of personal injury or death--applies in particular to fire departments and fire department members, but its underlying concepts apply to all types of emergency response operations.

**The goal: to reduce inherent risk**

Some risks are unavoidable and are accepted as part of the duty of a firefighter or a member of a public emergency response system. When one's mission is to save the property and lives of people in danger, one must expect to encounter danger. Emergency responders are recognized for their bravery because they are willing to accepts risks the general public finds personally unacceptable. Certainly, the nature of the activities of emergency responders exposes them to a high level of inherent risk--risk that all too often results in line-of-duty deaths and injuries.

Some risks are simply too great to take. Conditions or circumstances can exist in which the proper response is **not** to take action--action that could certainly place the lives of responders in danger. Training that prepares an emergency responder to recognize and respect dangerous situations and to work safely in a dangerous environment provides a considerable amount of counterbalance to the risk inherent in the work itself The equipment that makes it possible to survive and safely perform required actions also helps balance the risk.

**Identification, evaluation, and control**

Although we cannot always <u>eliminate</u> such danger from the work environment, we can <u>reduce</u> the risk of injury or death if we:
- know the nature of the *threat--Identification*
- determine the risk potential, based on signs and *symptoms--Evaluation*
- incorporate risk management practices into our plan for conducting operations-- *Control.*

*Continued on next page*

## Operational Risk Management, Continued

**New applications of accepted ideas**

The concepts of operational risk management have always been important considerations for command officers and supervisors in fire departments and emergency response organizations. Even so, the actual application of risk management as a focal point of incident management is a relatively new development.

**Operational risk management involves a higher level of risk**

A significantly higher level of risk management is involved in directing emergency operations and in regulating the overall exposure of responders to the risks of an incident. Emergency incident risk management expands the standard approach to directing and conducting emergency operations by incorporating risk evaluation and the assessment of optional approaches to the problems at hand. Doing that serves as the basis for determining what levels of risk are acceptable in different situations and guides incident commanders as they make strategic decisions. To deliver emergency services with an appropriate concern for the health and safety of the personnel who provide those services, an incident manager or incident commander must balance the approaches he or she takes.

# Planning to Manage Risk

---

**Introduction**

In 1987, the National Fire Protection Association (NFPA) adopted *NFPA 1500, Standard on Fire Department Occupational Safety and Health Program.* It was the first consensus standard to directly address many issues that are related to the avoidance of injuries, fatalities, and occupational illnesses emergency response personnel experience in the performance of their duties. Considered radical by some at the time, the principles that form the foundation of the standard have since been accepted and adopted widely throughout the North American fire service.

---

**NFPA 1500--
1992: The Risk
Management
Plan**

Revisions to. *NFPA* 1500 were adopted in 1992. The revised standard requires that a **written** plan--called a **Risk Management Plan**--be made part of a fire department's official policies and procedures.

A Risk Management Plan establishes fundamental policy. The Plan serves as documentation that risks have been identified and evaluated and that further, a reasonable control plan has been implemented and followed. The components of a Risk Management Plan required by *NFPA 1500* are:
- risk identification
- risk evaluation
- risk control techniques
- program evaluation and review

The elements of a Risk Management Plan, as outlined in *NFPA 1500,* are intended to apply to all aspects of a fire department's operations and activities. It is their application to emergency operations in particular, though, that makes the Risk Management Plan a significant advancement in risk management.

---

*Continued on next page*

# Planning to Manage Risk Continued

**NFPA 1500**    The following table summarizes the provisions of *NFPA 1500* that include Risk Management Plan requirements.

| NFPA 1500 2-2—Risk Management Plan | |
|---|---|
| **Paragraph** | **Requirement** |
| 2-2.1 | The fire department shall adopt an official written risk management plan that addresses all fire department policies and Procedures. |
| 2-2.2 | The risk management plan shall cover administration, facilities, training, vehicle operations, protective clothing and equipment, operations at emergency incidents, operations at non-emergency incidents, and other related activities. |
| 2-2.3 | The risk management plan shall include at least the following components:<br><br>Risk Identification: Potential problems<br>Risk Evaluation: Likelihood of occurrence of a given problem and the severity of its consequences<br>Risk Control Techniques: Solutions for elimination or mitigation of potential problems; implementation of the best solution<br>Risk Management Monitoring: Evaluation of effectiveness of risk control techniques. |

*Continued on next page*

# Planning to Manage Risk, Continued

**NFPA 1500 Appendix**

The following table summarizes the provisions of *NFPA 1500 Appendix* that include Risk Management Plan guidelines.

| NFPA 1500 Appendix—A2-2.3—Risk Management Plan | |
|---|---|
| **Paragraph A2-2.3** | **Guidelines** |
| | Risk Identification: List potential problems for every aspect of the operations of the fire department. The following sources of information should be useful in the process:<br>• a list of the risks to which members are or may be exposed<br>• records of accidents, illnesses and injuries, both locally and nationally.<br>• reports on inspections and surveys of facilities and apparatus<br>Risk Evaluation: Evaluate each item listed in the risk identification process, asking:<br>• What is the potential frequency of occurrence?<br>• What is the potential severity and expense of its occurrence?<br>Consult:<br>• safety audits and inspection reports<br>• accident, illness, and injury statistics<br>• applications of national data to local circumstances<br>• professional judgment in evaluating risks unique to the jurisdiction<br>Risk Control: Develop and implement strategies to:<br>• eliminate or avoid the activity<br>• reduce or control the risk<br>• develop, adopt, and enforce safety programs and standard operating procedures<br>• provide training<br>• conduct inspections<br>Risk Management Monitoring and Follow-up: Periodic evaluations should be made to determine how effectively the plan is working and identify modifications that should be made. |

# Chapter Summary

| | |
|---|---|
| **Risk management concepts and operational practice** | The application of risk management concepts to the operational practice of fire departments is the primary focus of this manual. Throughout the manual, we address specific challenges involved in incorporating a "Risk Management Approach" into the delivery of fire department services. It is a goal of this manual to assist departments and their leaders in their efforts to improve the performance of their organizations and to reduce the risks their communities, their organizations, and their individual members face virtually every day. |
| **Key points** | Key concepts covered in this Chapter include:

Levels of risk depend on the interrelationship of three factors:
- the probability that an undesired event might occur
- a harmful or undesirable consequence
- the severity of the harm that might result.

Individuals who provide emergency services accept a higher level of risk than members of the general public--hopefully, we can learn to manage them.

Risk management is a dynamic system that must be managed, not a "worry of the week" problem to be solved.

A significantly increased risk of injury or death is part of the environment in which emergency responders are expected to perform their duties.

It is impossible to avoid many of the risks inherent in an emergency responder's work environment, but, through control measures, we take steps to minimize those risks.

The risk of injury or death is reduced by control measures, including:
- training, experience, protective clothing and equipment
- implementation of appropriate strategies and tactics
- avoidance of unnecessary risks

*NFPA 1500, Standard on Fire Department Occupational Safety and Health Program* requires the establishment of a written Risk Management Plan. |

*Continued on next page*

# Chapter Summary, Continued

**Key points** (continued)

The application of risk management to emergency operations is intended to recognize the inherent risks and assist an incident commander in implementing appropriate plans and actions in dangerous situations. Development and implementation of any community's risk reduction efforts must address engineering, education, and enforcement programs.

Fire departments must manage financial, liability, and safety risks within three major categories:
- risk to the community--*Community risk*
- risk to the fire department organization--*Organizational risk*
- risk during emergency operations--*Operational risk.*

# Organizational Risk Management

# Chapter Overview

**Purpose**
The objectives of this Chapter are to:
- define risk management and loss control
- introduce the general principles of risk management and loss control as they apply to organizations that deliver emergency services
- present a five step risk management process that can be used by emergency service providers
- address legal responsibility as a component of risk management
- address specific areas of concern relating to risk management for fire departments and other emergency service providers.

**In this Unit**
This Chapter includes the following topics:

# Definitions and Concepts

**Risk**

The <u>Oxford American Dictionary</u> defines risk as:

1. The possibility of meeting danger or suffering harm or loss.
2. A person or thing insured or similarly representing a source of risk.
3. To expose to the chance of injury or loss, to accept the risk of.

*NFPA 1500* defines risk this way:
"A risk is a measure of the probability and severity of adverse effects." The adverse effects result from an exposure to some type of hazard.[1]

The concept of risk includes two dimensions of probability--the probability that an undesirable event will occur--and the probable magnitude or severity of the undesirable consequences. "Risk" is used in several contexts to refer to undesirable consequences that might occur in different situations. Those consequences might include:
- death
- injury
- property loss or damage
- increased operating costs
- payments for losses incurred by others
- loss of the ability to provide service
- inconvenience, and many other considerations.

**Management**

<u>Webster's New World Dictionary</u> defines management as:
"The act, art, or manner of managing or handling, controlling, directing".

"Management" suggests an organized and directed approach that implements evaluated techniques to control systems, events and people. Risk management implements a "proactive" rather than a "reactive" approach to solving problems or limiting risks. The term "risk management" refers to a systematic effort to <u>identify, evaluate and control risk(s)</u> to reduce both the probability that something might go wrong and the adverse effects (magnitude) if something <u>does</u> go wrong.

*Continued on next page*

---

[1] Reprinted *with* permission from *NFPA 1500, Standard on Fire Department Occupational Health Program,* copyright © 1992, National Fire Protection Association. Quincy MA 02169. This reprinted material is not the complete and official position of the National Fire Protection Association on the referenced subject, which is represented only by the standard in its entirety.

# Definitions and Concepts, Continued

**Exposure**

Exposure is an important term related to risk management. Exposure is a threat that some action or even non-action can lead to a loss of some kind. Recognizing and identifying exposures is an essential step in any risk management program.

**Risk management**

Risk management incorporates a full range of measures that may be used to limit, reduce or eliminate the probability that an undesirable outcome will occur. It also includes all types of measures that can be used to limit, reduce or eliminate the anticipated magnitude of the undesirable outcome, if it does occur. Risk management measures may address the probability of the occurrence, the probable magnitude of the outcome, or both.

Fire departments and other organizations that deliver emergency services are involved in many situations that could result in undesirable outcomes, including death or injury to members of the department while delivering emergency services. Other undesirable outcomes include loss or damage to the organization's apparatus, equipment or facilities that would have to be replaced or repaired. They would also include the death or injury of other persons that could result from errors or omissions by the emergency service providers, as well as damage to their property.

Managing organizational risk is not unique to fire departments. All organizations must manage some types of risk. The nature of the activities conducted by a fire department makes risk management a highly important and challenging task. Some potential exposure areas and specific examples within each of those areas are listed on Figure 1.

**A fire department's mission is risk management**

Managing risk for others is, in fact, the fire department's mission. The community is always at risk from an endless number of potential hazards or sources that could cause harm. The mission of the fire department is to reduce the probability of harm to the community that could result from different harmful situations and circumstances. The fire department must manage its internal (organizational) risk while it performs its mission of managing external (community) risk. This manual addresses managing risk within fire departments and other emergency service organizations.

*Continued on next page*

# Definitions and Concepts, Continued

**Supporting the mission statement**

Every fire department has the need to explicitly express its mission in the form of a mission statement, That formal document can be the starting point for a department's risk management efforts and can form the base from which to create a broad spectrum of risk management policies, processes, and procedures to be implemented throughout every level of the entire organization--top, middle, and bottom.

**High risks require special attention**

Different activities that are performed by emergency service organizations involve exposure to different kinds of risks. The primary mission of fire departments is to reduce the probability that the community will be damaged or destroyed by fire and the probability that deaths or injuries will result from fires. The probability of fire occurrence is addressed through fire prevention and public education activities, which are relatively low risk activities. Fire suppression and rescue functions are conducted to limit the damage and other negative consequences that result when a fire does occur and involves significantly higher risk to the service providers.

Examples:
- Emergency medical services reduce the risk of death or disability when citizens are injured or suffer from serious illnesses,
- Hazardous materials response teams address the risk that is created by harmful substances that have escaped from their normal containment.
- Confined space and technical rescue teams conduct high risk tasks that require the highest level of risk management when responding to incidents that other people or organizations are not prepared to manage.

Although each emergency service organization is associated with a dimension of risk management at the community level, this manual is directed toward management of *internal risk*--which includes all the things that can go wrong when departments attempt to deliver the services that define their mission. It also includes any undesirable outcomes that might result from non-emergency activities and other functions the organization performs.

*Continued on next page*

# Definitions and Concepts, Continued

**Exposures: risk management techniques**

Figure 1 provides examples of key areas of potential exposure and associated organizational risk management techniques.

| Potential Exposure Area | Risk Management Techniques |
|---|---|
| **Personnel** | |
| • Failure to meet minimum performance requirements | • Establish minimum performances |
| • Failure to properly train | • Establish and conduct performance-based training for all personnel |
| | • Training should conform to relevant OSHA, NFPA, and other standards |
| • Failure to adequately equip | • Provide protective equipment that meets NFPA standards. |
| **Fire Inspection Practices** | |
| • Failure to notify owners of hazards | • Require complete records of every inspection |
| • Failure to pursue compliance | • Consistently issue citations and seek judicial intervention when hazards are imminent |
| **Administration** | |
| • Level of service not defined | • Define level of service for all service deliverables |
| • Incomplete records | • Document and address all complaints promptly (policy) |
| **Communication** | |
| • Failure to dispatch promptly | • Ensure that specific dispatch policies are in place and that performance is monitored |
| • Failure to properly advise callers of potential delay | • Establish policy and procedure to address these issues |

**Figure 1: Exposures and Techniques**

*Continued on next page*

**Delivering emergency services involves high risk activities**

Every person and every organization has to face a variety of risks in life.

Examples:
The risk of being involved in a traffic accident, or being injured in an accident.

The risk of your house being struck by lightning.

The risk of your being struck by lightning.

The risk of a floor collapse or the risk of being on the floor when it does collapse.

Everyone also participates in some forms of personal and organizational risk management, such as driving safely to avoid accidents and wearing seat belts to limit injuries if an accident should occur. Fire departments and emergency service organizations have a special need for risk management because:
- they engage in high risk activities that expose them to elevated risk.
- they are responsible for delivering critical emergency services to protect their communities.

If we didn't attempt to manage risk, we would be leaving our fate to chance. There is a possibility we would soon be dead, injured, battered, broken-down or incapable of protecting anything. The risk management process is a system developed to limit both the probability that undesirable events will occur and the magnitude of those consequences if they do occur. We will never be able to eliminate all risks, but we can eliminate some, reduce many, and limit even more.

Examples:
To reduce the probability of firefighters being injured, we develop and conduct training programs. Driver training reduces the probability of vehicle accidents.

To reduce the magnitude of injury if a vehicle is involved in an accident, we enforce a policy that requires the use of seat belts.

# Losses and Loss Control

**Loss control is a component of risk management**

Loss control is a component of risk management, The objective of loss control is to limit the consequences of risk. There are hundreds of potential losses that are faced by fire departments, but essentially they fit into four major categories:

1. **Personnel Losses.** These losses include life loss, injury, and illness to members of the department. People are important; they are our most important asset and they must be protected.
2. **Property Losses.** Fire department property includes vehicles, facilities and equipment. Losses in this category could include; an apparatus that is damaged or destroyed in an accident, a fire station that is damaged by a fire, or a piece of equipment that is lost or stolen.
3. **Down time Losses.** These losses result when property is not available for it's intended use. An apparatus that is out of service as the result of an accident is not available to respond. A fire station that is damaged by a fire may not be able to house firefighters or apparatus.
4. **Liability Losses.** Liability is the obligation to compensate others for losses and damages that are caused by our acts or omissions. The frequency and magnitude of this type of loss for fire departments are rapidly increasing. A person who sues the fire department for negligence after being injured in an accident with a fire truck creates a significant potential liability to the department.

Most losses can be related to some form of cost to the organization, Loss reduction is intended to minimize these costs, which can be classified as direct or indirect.

*Continued on next page*

**Direct and
indirect losses**

Life safety--that of the public and of our response team members--is always the primary concern. The cost of loss can be high for individuals as well as organizations.

Direct costs are usually easy to identify. The loss of a piece of equipment will cost a certain number of dollars to replace. An apparatus involved in an incident will cost "X" dollars to repair.

Indirect costs associated with these losses, such as the impact of not having that piece of equipment available when it is needed might be difficult to determine. There are many types of indirect costs that should be recognized.

For example, an injury to a firefighter often costs much more than the direct costs associated with treating the injury. Direct costs would involve doctors' bills and hospital expenses. Worker's compensation insurance might pay for that firefighter's lost time and medical treatment, but higher claims could cause the costs of insurance to increase. The injured firefighter's position might have to be filled using overtime. If the firefighter is permanently disabled, the department might have to recruit and retrain a replacement. The firefighter's experience and knowledge might never be replaced. The department might also have to pay for a doctor's evaluation of the firefighter's medical condition and for a lawyer to process a disability pension.

# Organizational Risk Management System

| | |
|---|---|
| **Introduction** | The fire department risk management program is a system of functions, components, and activities designed to reduce the level of risk throughout the organization, All department members are responsible for managing various components of that system. |

| | |
|---|---|
| **NFPA 1500 mandates planning to manage organizational risk** | Section 2-3.1 of the National Fire Protection Association's *NFPA 1500, Standard on Fire Department Occupational Safety and Health Program* states:<br>*"The fire department shall adopt an official written departmental occupational safety and health policy that identifies specific goals and safety objectives for the prevention of accidents and occupational injuries, illnesses and fatalities."*<br><br>*NFPA 1500's* intent is to provide the framework for a safety and health program for a fire department or any other organization that provides similar services. |

| | |
|---|---|
| **No single solution or method** | There is no single method or solution to manage risk. Numerous publications provide a wide range of suggested methodologies for managing risk. The Appendix references many of those publications. Determining how to manage risk is a decision each department must make individually. The process must be frequently reviewed and, where necessary, upgraded. Most importantly, once a risk management process is instituted, it must be properly managed, continuously evaluated, and updated at least annually. |

*Continued on next page*

# Organizational Risk Management System, Continued

**The role of a risk manager**

In <u>Emergency Incident Risk Management, A Safety & Health Perspective,</u> Murrey E. Loflin and Jonathan D. Kipp offer this profile of the Risk Manager:

"This individual will typically have oversight over the risk management programs of several departments, emergency services among them. However, many of the functions can be centralized, which will relieve the department's administrator from some of the risk management tasks. Most frequently, the risk manager will handle relations with outside agencies such as insurance companies and/or agents, and can be responsible for handling the insurance needs of various departments. In addition, this individual is familiar with the overall risk management process, and can serve as a valuable resource for guidance and information."[2]

*Continued on next page*

---

[2] Kipp, Jonathan D. and Loflin, Murrey E., <u>Emergency Incident Risk Management. A Safety & Health Perspective,</u> 1996.

**Goals and objectives: the foundation of the system**

A risk management system is a dynamic process. To be effective, it should begin with the establishment of goals, followed by the establishment of objectives to meet those goals. As in any system, the goals and objectives should be:

- clearly stated
- understood
- attainable
- measurable.

Those who are expected to see that the goals are met should be involved in the process.

Risk management should be looked upon as a continuous, on-going, standard process for an emergency service organization. There will always be risks, and there will always be room for improvement.

One of the fire department's goals should be to minimize injuries. For example, the objective might be established as:

"Limit on-the-job injuries to zero per year."

To meet an objective, the department should establish specific plans or actions that support its achievement. One action might be that all department members participate in a training program designed to increase member awareness of the causes of injuries. Another might be to ensure adequate lighting is provided at incident scenes so members can identify and avoid safety hazards.

*Continued on next page*

**Fundamental steps: identification, evaluation, and control**

Any system for managing risk must provide for three fundamental steps:
- identification of risk
- evaluation of the probability and the potential magnitude of losses related to those risks
- the establishment of appropriate control measures.

Agencies, organizations, and individuals have developed several different models that can be used as a foundation for a fire department risk management program. Most of the models use five to ten steps. The model presented in this text uses five steps.

The essential steps of all of the models include the identification of risks and the evaluation of the potential harm that can result. That is followed by the identification, selection and implementation of control measures. The number of steps depends on how these individual core steps are broken down.

No matter how many control measures are applied, it must be recognized that losses will most likely occur and that control mechanisms for funding those losses must also be provided.

# The Steps in the Process

**Five principal risk management steps**

The five steps illustrated below provide a solid foundation for developing a risk management program.

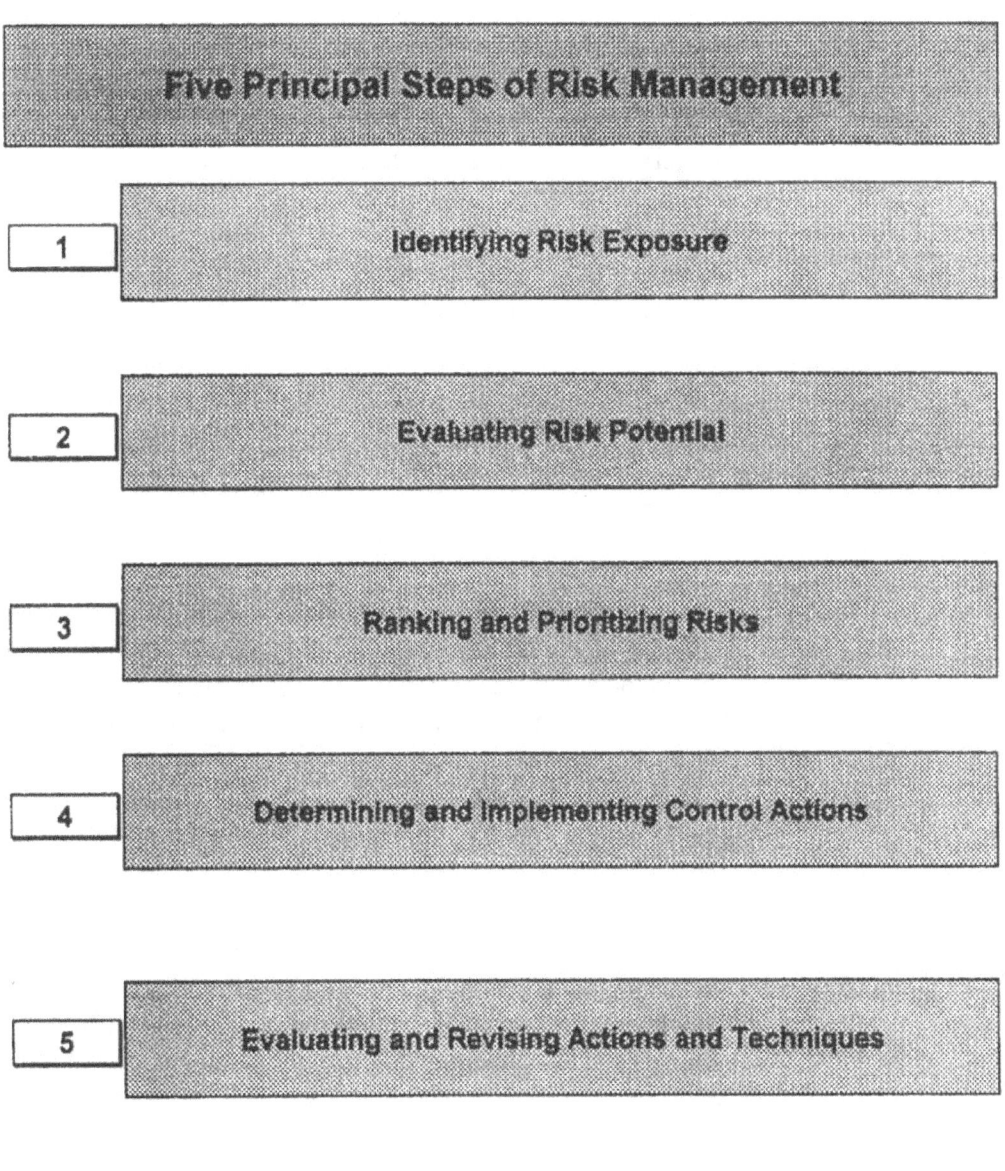

Figure 2: Five Principal Steps of Risk Management

*Continued on next page*

# The Steps in the Process, Continued

**Step 1:**
**Identifying risk**
**exposure**

This is the foundation of an integrated process for managing risks. Its purpose is to identify the kinds of things that create risks to the fire department. A fire department's exposure to risk stretches beyond emergency response and training activities. Potential risks in all activities can be grouped in several different ways. General areas of risk to consider include:

**People:** deaths, illnesses and injuries, health exposures.

**Apparatus and vehicles:** accidents, malicious acts, damage due to mechanical failure, operator error.

**Occupancies/Facilities:** natural disasters, fires, malicious acts, failure to open apparatus bay doors before driving out.

**Equipment:** theft, damage from use, damage from misuse, failure to close doors before leaving.

Look to your own department first. Use incident, injury, accident and other internal reports to identify local experience and trends.

Potential risks should also be identifiedthrough reading articles and reports and analyzing data from other sources of information. Neighboring departments that are "similar" might have experienced losses that can alert you to specific potential losses. Comparing data with other organizations or national averages could also identify problems that might not have been recognized within the organization. A fire department that records one injury per fire might consider that rate routine until it finds that surrounding fire departments have less than one injury for every five fires.

Numerous professional and trade organizations can provide relevant risk identification information: The International Association of Fire Fighters, the International Association of Fire Chiefs, the National Fire Protection Association, and the Risk and Insurance Management Society all provide members with observations, considerations and recommendations related to risk.

The Appendix provides a list of organizations and their addresses that can provide valuable information.

*Continued on next page*

---

**Step 2:**
**Evaluating risk**
**potential**

The evaluation of risk potential involves determining or estimating the likelihood that an event will occur and the consequences that will result if it does. Probability is generally established by studying the frequency at which incidents have occurred in the past. However, an undesirable event might not have a local history. The fact that something has not happened to one particular fire department in the past does not ensure that it will not happen in the future. Similarly, the magnitude of a loss might be predictable from past experiences, but it is important to remember that **incidents with the most severe consequences are usually the most rare.**

Note.
If incidents with severe consequences occurred more often, we would be compelled to prevent them. This is a simple illustration of the principles of risk management.

The evaluation step should provide information to answer the following questions:

1. What is our local experience?
2. What do we know about national experience?
3. What are the probabilities of different things happening?
4. What are the probable consequences if they do occur?

In evaluating risk potential, both the likelihood and loss potential need to be addressed simultaneously.

In the book mentioned previously, Emergency Incident Risk Management, A Safety & Health Perspective,[3] Kipp and Loflin refer to using "frequency" and "severity" as measures to evaluate risk potential. The text provides a methodology and Sample Risk Management Plan that serve as effective tools for integrating frequency and severity factors into a document with which to put a department's risks into priority.

*Continued on next page*

---

[3]Kipp, Jonathan D. and Loflin, Murrey E., Emergency Incident Risk Management, A Safety & Health Perspective, 1996.

**Step 2: Evaluating risk potential** (continued)

Frequency is an estimate of how often a loss has occurred--or is likely to occur--as a result of any of the risks determined during the identification step. Severity, which constitutes the second half of the evaluation phase, estimates the potential losses to the organization posed by the identified risk.

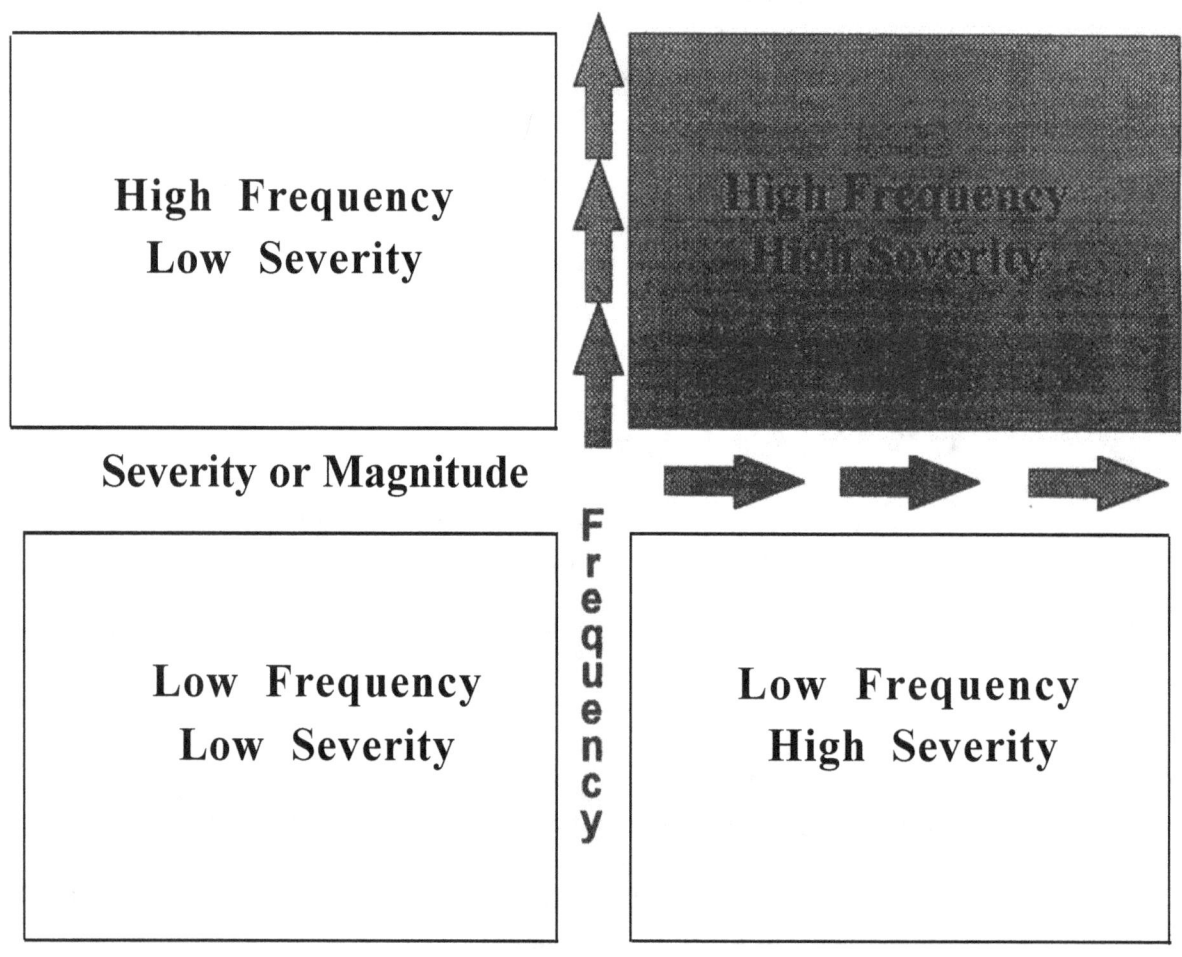

Figure 3: Establishing Priorities for Risk Management Actions

*Continued on next page*

---

**Step 3:**
**Ranking and**
**prioritizing**
**risks**

After considering the probabilities of occurrence and the probable outcomes, the next step is to prioritize the risks and decide on the areas that need to be addressed as priorities.

Generally, risks with the most severe potential outcomes are considered ahead of risks with relatively minor outcomes. Those that are more likely to occur are prioritized ahead of those that are less likely to occur. Doing that requires a considerable amount of judgment and a thorough analysis of the practicality of addressing certain types of risks.

Most organizations will be able to identify a fairly long list of risks that deserve attention, They should be able to address several of them simultaneously or in fairly rapid succession--including some that are relatively easy and inexpensive and some that might require a major effort over a lengthy period of time. The compilation of a prioritized list of areas needing attention is an important step in managing risk.

Example:
The application of a band of reflective tape to the front, rear and sides of all emergency vehicles to increase their visibility can be accomplished quickly and inexpensively. On the other hand, the replacement or modification of existing apparatus to provide fully enclosed seating areas for all crew members might take several years and involve a considerable expense. Enforcing a policy that requires all firefighters to remain seated with their seat belts fastened while vehicles are in motion should not involve any expense or take more than a few hours to implement.

---

*Continued on next page*

**Step 4:
Determination
and
implementation
of control
actions**

Several factors must be considered in determining control actions. Those factors are often interrelated; they will frequently make it difficult to act on one without having an impact on another. Before implementing control measures, the cost and associated benefits will have to be considered. The benefits will frequently deal with improved safety and health to personnel-- even so, if they cost real dollars, someone will have to justify the cost. In determining which control actions to implement, the following factors should be considered:

- predicted effect
- time required
- time to results
- effort required
- associated costs
- insurance costs
- expense funding
- cost/benefit
- mandated.

*Continued on next page*

## The Steps in the Process, Continued

**Factors to be considered**

- **Predicted Effect:** What savings will likely result? This factor will require the risk manager to estimate the predicted effect of control actions that would be implemented. When considered in conjunction with the cost to implement the control, it will determine the cost/benefit factor.

- **Time Required:** How long will it take to implement the control measures? Can the resources needed to control the risk be used more efficiently and effectively during that time? Will other efforts be adversely affected?

- **Time to Results:** This is not the same as time required. Long term results are a very difficult "sell" in today's environment. The community and those responsible for allocating its tax dollars might not be patient. If the effort is critical to the department's risk management goals and results will not be seen for an extended time, effective communication of the plan and its benefits to the community, its leaders, and its elected officials, is essential.

- **Effort Required:** How much is required and can that effort be more effectively applied to other programs? Is there more than one solution? Will one solution require less effort than others? If you're looking for more efficient ways to address a risk, involve the people it affects the most,

- **Associated Costs:** How much will it cost--directly and indirectly? How do you arrive at and document those costs? Cost alone will often determine if a proposed control measure is implemented or ignored. It will always affect where that measure appears on a priority list. Although it is a component of the cost/benefit factor, it will also be viewed separately. As discussed earlier, all factors need to be considered collectively. The determination to implement control measures directly related to the safety or health of workers must not be made based solely on cost. The risk manager, however, must be able to properly establish, and then communicate both direct and indirect costs.

*Continued on next page*

---

**Factors to be considered** (continued)

- **Insurance Costs:** Will initiating the control measure reduce the related insurance cost? Will to implement it increase that cost? If so, is the cost reduction or cost increase worth instituting or not instituting the measure? Costs for almost all insurance programs are determined by viewing loss experiences generically and also on a customer specific basis. Costs are established by estimating anticipated losses. Departments with high accident or injury rates will pay more. Those with less will pay less.

---

**Funding**

Other Funding Mechanisms: Risk management requires a balance of other risk control and risk financing measures. The measures or methods used should support each other. They are essential to any risk management strategy. Effective risk control efforts will limit the costs of risk financing while financing will limit address costs when control measures fail. There are two principal types of risk financing:
- risk retention
- risk transfer.

Risk retention relies on internal funds. Risk transfer, on the other hand, looks outside the organization for money needed to pay for losses

Risk retention techniques include:
- budgeting or expensing to pay for relatively inexpensive or small losses such as minor damage to apparatus, equipment or facilities
- establishing a reserve fund to address irregular losses
- borrowing funds to pay for unexpected losses. In this case, the borrower has an obligation to pay a specific amount for a given loss. The risk is therefore retained and not transferred.

Risk transfer techniques include:
- purchasing commercial insurance and contractually transferring the risk to another entity.
- insurance that transfers the financial burden for future risks to the insurer in exchange for a fee
- contractual transfers that involve another entity's assuming responsibility for the financial burden of a loss. This option is frequently referred to as an indemnity clause.

---

*Continued on next page*

**Cost/benefit analysis**

Ranking and prioritizing risks is an integrated process that requires that the frequency and severity of a risk be quantified and ranked in some way. Each of the ranked risks should then be prioritized by evaluating a series of risk factors. The risk manager will seldom have all the data required to make a quantitative assessment and will have to rely on knowledge, experience and judgment to make required assumptions.

The risk manager will need to consider all the factors above in developing a cost/benefit analysis. Although the benefits will frequently deal with improved safety and health of personnel, if they cost real dollars, someone will have to prepare a "balance sheet" (cost/benefit analysis). The sheet will require real and valid numbers. Someone will have to justify the numbers and prove the benefits are worth the investment. Figure 4 presents a sample "balance sheet" for conducting a cost/benefit analysis. Many of the publications referenced in the Appendix also provide recommendations or models for conducting a cost/benefit analysis. Figure 4 is intended to be a tool, and like any other tool, can be modified. Use a tool that meets your needs. No matter how you conduct a cost/benefit analysis, the value of your findings will depend upon the quality of your data and your assumptions.

*Continued on next page*

# The Steps in the Process, Continued

## Cost/benefit analysis (continued)

| | |
|---|---|
| **ESTIMATED SAVINGS AFTER CONTROL COSTS:** | |
| • Total Estimated Savings | **$7160** |
| • Total Cost of Control | **$ 325** |

**I. CURRENT STATE:** (Briefly describe assumptions and provide facts)

    **Example:**    There were 25 hand injuries that resulted from packing hose during the previous year.

• **Direct Costs:** (describe assumptions or provide costs)
    1. Hospital/Medical Costs - 25 X $250 = $6250
    **TOTAL DIRECT COSTS**    **$6250**

• **Indirect Costs:** (describe assumptions or provide costs)
    1. Additional Staffing to Fill Positions - 15 shifts x 180 per shift = $2700
    **TOTAL INDIRECT COST**    **$2700**
    **TOTAL CURRENT STATE COSTS**    **$8950**

**II. FUTURE STATE:** (describe assumptions or provide facts)

    **Example:**  Revised procedure and training will prevent 80% of all identified injuries

• **Direct Costs:** (describe assumptions)
    1. Hospital and Medical Costs - 5 x $250 = $1250
    **TOTAL DIRECT COSTS**    **$1250**

• **Indirect Costs:** (describe assumptions)
    1. Additional Staffing to Fill Positions - 3 x $180 = $540
    **TOTAL INDIRECT COSTS**    **$ 540**

    **TOTAL FUTURE STATE COSTS**    **$1790**

**III. TOTAL ESTIMATED SAVINGS PRIOR TO CONTROL EXPENDITURES:**
    • Total Cost Current State = $8950
    • Total Cost Future State = $1790
    • **TOTAL ESTIMATED SAVINGS PRIOR TO CONTROL EXPENDITURES**

**IV. COSTS ASSOCIATED WITH CONTROL MEASURE** (briefly describe)
    **Example:** Revised operational procedure and shift training will reduce injuries by 80%

• **Direct Costs:** (list the direct costs associated with the proposed control measure)
    1. Cost of Video = $325
    **TOTAL DIRECT COSTS**    **$ 325**

• **Indirect Costs:** (list the direct costs associated with the proposed control measure)
    1. Training - Shift Requirement - No Cost
    **TOTAL INDIRECT COSTS**    **$ 0**

    **TOTAL COSTS ASSOCIATED WITH CONTROL MEASURE:**    **$ 325**

**V. TOTAL**    **TOTAL ESTIMATED SAVINGS AFTER CONTROL COSTS**    **$6835**

**Figure 4: Example of a Cost/Benefit Analysis Balance Sheet**

*Continued on next page*

# The Steps in the Process, Continued

**Options and categories of control measures**

The means by which potential risks are addressed are referred to as *risk control options.* Methods that are used to address the risks are categorized into three *risk control categories.* The specific steps that are taken to control the risks are referred to as *risk control measures.* In almost all cases, the options, categories and specific methods selected are intended to limit the risk potential.

There are three risk control options:
- risk avoidance
- risk reduction
- risk transfer.

**Risk control options**

**Risk Avoidance:**
Total risk avoidance is a sure route to reducing risks. It is usually the safest solution but often the least realistic. Many times, it is no option at all. We can't stop going to fires and all fires present some risks. If an action, activity, or condition poses a risk to the operation, consider whether or not it can be avoided or eliminated entirely. Although we are required to respond to a structural fire, we can choose not to enter the structure.

**Risk Reduction:**
Risk can often be minimized by planning, training, testing, maintenance of standards, and enforcement of codes. The fundamental purpose of standard operating procedures, planning, and training is to reduce the risk to personnel. Proper selection, design, and maintenance of facilities and vehicles is intended to reduce various forms of risk.

*Continued on next page*

---

**Risk control options** (continued)

### Risk Transfer:

It may be feasible to transfer our risks to other parties by purchasing insurance or contracting out to other organizations to perform certain activities. Good risk management practices should reduce insurance costs. In addition to assuming a portion of the risk, many insurance companies provide valuable assistance in identifying and reducing risks. By doing so, they intend to reduce a department's losses and the probability of claims.

---

**Risk control categories**

When it is not possible to eliminate risks, they should be limited or minimized. For example, although we can't eliminate the need for firefighters to respond to emergencies, we can minimize the risk they face while responding to and operating at an incident.

Risks may be systematically eliminated or minimized by implementing control measures. Control measures fit into three general areas which we call Risk Control Categories. The general categories of control are:
- administrative
- engineering
- personal protection.

Administrative controls, such as standard operating procedures, and engineering controls, such as apparatus specifications, are directed toward making the workplace safe for the firefighter. In contrast, personal protection controls such as self contained breathing apparatus are focused on making the firefighter safe for the workplace. When establishing a risk control program, fire departments should use administrative and engineering controls to limit exposure to risk before relying on personal protection control measures.

---

*Continued on next page*

# The Steps in the Process, Continued

**Three categories of risk control**

**Administrative Controls:**
Administrative controls provide the foundation for a department's risk control program. They include:
- emergency vehicle operation procedures and regulations
- training requirements
- enforcement of fire codes to eliminate hazards
- pre-fire planning to identify hazards
- accountability systems.

Consistent application and enforcement of administrative controls constitute an essential administrative control. If you have a policy and don't enforce it or a required procedure and don't require it, in reality you have neither a policy nor a procedure.

**Engineering Controls:**
Engineering controls are intended to remove specific hazards from the workplace. Fire departments use engineering controls to improve safety and health in fire stations and on apparatus by designing improved components or entire systems. Improving ventilation at a fire station to remove diesel exhaust or relocating the siren on an apparatus to reduce noise in the cab, providing foam for flammable liquid fires and lighting for night operations are examples of engineering controls.

**Personal Protection Controls:**
Personal protection controls do not remove hazards from the workplace. They are designed to provide an element of personal safety. Personal protective clothing, self-contained breathing apparatus, PASS devices, and life safety rope are examples of personal protection controls. When operating at emergency incidents, firefighters are required to rely heavily on personal protection controls.

*Continued on next page*

---

**Step 5:**
**Evaluate and**
**revise actions**
**and techniques**

Risk management should be a continual process with established mechanisms to monitor performance and evaluate progress, Risk management efforts should yield positive results in terms of better outcomes, The change may be sudden and obvious or it may take a long time to yield measurable results.

The evaluation stage should parallel the steps that were taken to identify risks and seek confirmation that the process is working. It should focus on the areas that were identified as requiring attention and provide follow-up to determine if they actually result in the desired change in outcomes. In many cases, this will include an improved focus on the problem areas, since a risk management action plan can include a component to more closely monitor the specific area of concern.

For example, if hand injuries were identified as a problem and new gloves were issued as the solution, the frequency and severity of hand injuries should be monitored to confirm that the desired impact is being achieved. Doing that might include a more detailed process for categorizing hand injuries and considering the effectiveness of the new gloves.

All actions should be evaluated routinely to ensure that they are achieving the desired outcomes and are not creating other unanticipated problems. Risk management is a process of balancing different factors. The balance can often be influenced by any number of changing circumstances.

---

**Risk**
**management is**
**a system, not a**
**solution**

Kipp and Loflin define the risk management process as: "...A system for treating pure risk; identification analysis of exposures, selection of appropriate risk management techniques to handle exposures, implementation of chosen techniques and monitoring of results."[4]

The risk management process is intended to minimize losses. It is a dynamic system and not a fixed solution. Most losses are avoidable costs, and no department can afford to pay for avoidable costs. The results of administrative, engineering and personal protection control systems will determine the success of your risk management program.

---

[4]Kipp, Jonathan D. and Loflin, Murrey E., Emergency Incident Risk management. A Safety & Health Perspective, 1996.

# Legal Responsibility

**Introduction**

The risk that an individual or organization could be held legally responsible for an undesirable outcome is one important consideration in risk management.

Organizations are generally considered to be legally responsible for any harm that results from their acts or omissions, and are expected to conduct their activities in a responsible manner that does not expose individuals or the community to an unreasonable level of risk. That basic principle could apply to a wide variety of "injuries", including:
- physical injury or fatality
- damage to public or private property
- damage to the environment
- negative economic impacts on an individual or corporation
- damage to an individual's reputation.

**Statutory immunity**

Because of the nature of the mission, many fire departments have limited statutory immunity under state laws. Statutory immunity protects them from. being held responsible for failure to provide adequate protection to save lives or property from fires. That assumes the fire department did not cause the fire and did not take any unreasonable actions that resulted in a greater loss.

Most legal decisions have found that the duty of a fire department is to protect the community at large from fires, which does not include a specific duty to protect individual citizens or their property. That gives the fire chief or the incident commander a large measure of authority to make decisions concerning whether it is reasonable or unreasonable to attempt to save property--based on the risks to fire fighters that would be involved in attempting to save it.

This type of statutory immunity is usually limited to the delivery of specific governmental services and may not apply to incident types other than fires. Depending on the state and location, contracted companies and independent volunteer organizations might not have the same statutory immunity. The rules are different in different states and for different services, particularly for emergency medical services--who have a direct service delivery relationship with their patients.

*Continued on next page*

**Responsibility
to provide a
safe and healthy
workplace**

The fire department, like other public or private organizations, also has legal responsibilities to manage the level of risk to which its members and employees are exposed in the performance of their activities. These duties are defined by occupational safety and health laws and regulations as an employer's responsibility to provide a reasonably safe and healthy work environment for its employees.

Virtually every human activity involves some degree of risk, and every occupation involves at least some potential for injury, illness, or even death. In most work relationships, the employer is considered to have almost unlimited responsibility to manage the level of risk. There are few occupations today where the willingness to accept a high risk of injury or death is considered to be an important qualification.

Because they cannot be held responsible for hazards that might be present where their work must be performed, fire departments and other emergency response agencies have a relationship to occupational risk different from any other type of organization. The nature of the work requires emergency responders to accept situations as they are found and to deal with whatever unplanned and uncontrolled event might ensue.

Although the risks involved in a specific situation might be beyond the ability of the responding agency to predict or control, the nature of the risks is usually predictable. Many risks are avoidable. Under those circumstances, fire departments must be prepared to work as safely as is reasonably possible.

Figure 5 summarizes some key factors about the fire department's work environment that must be considered. It is important to note that these factors apply only to the emergency response environment. The fire department has the same responsibility as any other organization for the safety of its members when they are working at a facility that is normally under the control of the department.

*Continued on next page*

---

**Responsibility to provide a safe and healthy workplace** (continued)

A fire department has extremely limited, if any, control over the emergency location until it is called to respond to an emergency.

The work environment is not designed with the expectation that the fire department might have to respond to work.

The nature of the work is dealing with situations that are too dangerous for anyone else to handle.

The urgency of the situation usually does not allow a fire department to the correct hazards that are present before they take action.

**Figure 5: Factors That Affect a Safe and Healthy Workplace**

---

*Continued on next page*

**Acceptance of risk**

Fire fighters and other emergency responders knowingly accept the increased risk of accidents, injuries, and potential death that is inherent in their work. They willingly operate in an elevated risk environment. In most states, legal precedents protect a property owner against negligence suits brought by firefighters for injury or death resulting from an incident on their property. What is the rationale behind such precedents? It is assumed that the firefighter was:

- aware of the risks involved in the activity
- knowingly accepted them.

That principle has been challenged in cases in which the evidence indicates the occupant or property owner was doing something that posed an unreasonable risk to the responders. The property owner's protection might be compromised if hazards have been knowingly concealed or if the owner has failed to correct fire code violations.

The requirement in *NFPA 1500* to incorporate risk management in the process of conducting emergency operations can be interpreted as increasing the duty of fire officers to be responsible for the health and safety of the personnel they command and supervise. That responsibility is different in some respects from the duty to provide protection to the community.

*Continued on next page*

**An incident commander must have responsibility WITH authority**

Because the primary responsibility of a fire department is to protect the community from fires, state and local laws usually provide a fire chief virtually unlimited power to control and determine the fate of private property that is burning or threatened by fire. An incident commander can "write off" a burning building by making the decision that the property cannot be saved without excessive risk to the firefighters. In an extreme situation, the decision could be to discontinue rescue operations because the slim possibility of a successful rescue does not justify the risk to rescuers. The laws provide for such discretionary power because it would be virtually impossible to conduct emergency operations under the threat of being legally challenged over every discretionary decision.

A fire chief must have the authority to make discretionary decisions for the overall good and public safety of the community. In the absence of gross negligence, the law provides a wide margin of protection for fire departments and incident commanders.

**Protection from liability**

It is within the legal rights of virtually anyone to file a suit against any individual or organization for anything bad that happens to them There are no guarantees that a fire department or fire chief will not be sued. The immunity principles make it easier to defend against such suits and often to have them dismissed as groundless. Still, it should not comes as a surprise to any public official to become a defendant in a legal action, *NFPA 1500* establishes a set of risk management principles that could be cited to justify a decision not to expose firefighters to excessive risks.

**Occupational safety and health regulations**

OSHA (the Occupational Safety and Health Administration of United States Department of Labor) enforces federal standards that apply to private sector workers and employees of the Federal Government. The primary regulations that apply to fire departments are OSHA's 1910.120 (Hazardous Materials), 1910.134 (Respiratory Protection), 1910.154, 155, and 156 (Fire Brigades), 1910.1030 (Bloodborne Pathogens), and 1910.1200 (Hazard Communication). Hazardous materials regulations are also enforced by the Environmental Protection Agency where OSHA does not have jurisdiction. In addition, every state has a regulatory structure to provide for the safety and health of workers.

*Continued on next page*

**"OSHA Plan States"**

Approximately twenty five states and territories (known as "OSHA Plan States") have agreements with the Department of Labor to enforce <u>federal</u> standards through State agencies. Even though the Department of Labor does not have jurisdiction over state and local government agencies, designated state agencies in the "OSHA Plan States" are required to enforce federal regulations on public agencies.

**Federal regulations establish minimum requirements**

OSHA regulations establish a minimum standard. Individual "OSHA Plan States'! may adopt equivalent or more stringent regulations. Those states also determine if they will enforce regulations for volunteer fire departments and other emergency response organizations, or if they will apply them to paid workers only. Increasingly, the trend is to enforce the same regulations on volunteer organizations and fully-paid departments.

**Other states**

OSHA itself enforces federal regulations in the remaining states, States that do not have agreements to enforce the federal regulations generally adopt their own regulations and determine how to apply them. For example, in some states, the regulation of fire departments is assigned--not to an occupational safety and health agency--but to the state fire marshal or a state agency responsible for fire protection regulations.

Policies and enforcement programs differ significantly from state to state, so it is important that each fire department and emergency response agency become familiar with specific regulations that apply and the state agency that has the authority to enforce them.

**A regulatory framework**

OSHA regulations and other federal and state regulations that apply to worker health and safety do not specifically address operational risk management. However, they do establish a regulatory framework that is intended to establish a safe working environment. In most cases, the regulatory language does not address the issue of an inherently dangerous occupation, but does hold an employer responsible for recognizing hazards and taking appropriate action to eliminate or minimize the risk of harm to the employees.

*Continued on next page*

| | |
|---|---|
| **Consensus standards** | Where there is no specific regulation, most regulatory agencies have the ability to refer to a consensus standard that addresses a specific issue. An agency may also refer to an established consensus standard that establishes a reasonable standard of care if it is more specifically applicable to a subject or situation than the regulation that is in force. |

In this manner, *NFPA 1500* establishes a basic foundation for a risk management approach to emergency operations. A fire department that adopted and conscientiously followed the requirements of *NFPA 1500* would be in a very good position to show that is approaching risk management proactively and responsibly. The same basic concept applies to all of the provisions of *NEPA 1500*. A fire department that adopts and follows the requirements of *NFPA 1500* will meet or exceed the requirements enforced by most regulatory agencies.

It is very difficult to determine if a fire department has an effective approach to operational risk management without examining its application to specific situations. A regulatory agency might examine a department's standard operating procedures and training programs to see what is documented, but it is difficult to judge from the written procedures if a risk management approach is routinely applied and if its application is reasonable.

Actions taken at a particular incident might be reviewed by a regulatory agency to determine if appropriate procedures were followed.

Example:
An aerial ladder collapses at a fire, causing serious injury to a firefighter. Was the ladder tested by a qualified individual or company within the previous year? Was it regularly inspected for defects? Was the operator properly trained?

Example:
A firefighter becomes lost in a burning building, runs out of air, and dies. Was his or her SCBA properly maintained and tested? Was the air quality checked? Was the firefighter wearing a PASS? Did they use a buddy system? Was accountability used? Was there an appropriate incident command system in place?

*Continued on next page*

**Consensus standards** (continued)

Example:

A fire in a chemical facility results in a major ecological incident when water runoff carries contaminants into a protected wildlife refuge. Did the fire department take appropriate action to control or prevent the runoff? Did the effort to control and extinguish the fire create a greater problem than letting it burn? Is the fire department or the owner of the facility responsible?

In such cases, it is likely that the risk management approach defined in *NFPA 1500* would be used as a yardstick to determine if appropriate judgment was used. There have been relatively few cases in which a fire department has been fined or formal warnings have been issued for failure to comply with occupational safety and health regulations at particular incidents--although substantial fines have been assessed in some of those cases.[5] There are no known cases in which a fire department has been fined specifically for failure to exercise appropriate risk management.

**Documentation is essential**

Because operational risk management requires a large measure of subjective judgment, its performance can be most easily evaluated after an incident occurs. Risk management practices are most likely to be reviewed in two sets of circumstances:
- if a regulatory investigation is prompted by an accident that has resulted in a death or injury
- if a complaint is filed by a member of the department.

*Continued on next page*

---

[5]Enforcement actions against fire departments are generally initiated by state occupational safety and health authorities, which have jurisdiction over local government agencies.

# Legal Responsibility, Continued

**Documentation is essential** (continued)

A fire department should be able to demonstrate that its individual officers and members:
- are competent and well trained in the duties they are expected to perform
- are familiar with the regulations and current standards that apply to those duties
- follow established standard operating procedures and guidelines that are consistent with the regulations and standards.

The department should also be able to show that it applies a consistent approach to risk management to every incident.

# Specific Areas of Concern for the Emergency Service Risk Manager

**Introduction**    The responsibility for managing risk in an organization that exists to provide emergency public safety services is extremely challenging. On one side of the equation, it can be easy to categorize many risks as inherent, and to let the objective--saving lives and property--dominate the mission. On the other side of the equation, an approach that is too conservative could render the service ineffective.

In some situations, poor risk management judgment can create even greater risks. For example, a fire department that arrives too late will be unsuccessful at saving lives or limiting property losses, but a fire department that drives too fast in order to arrive quickly may cause as many deaths as it prevents.

The only reasonable approach is to embrace a professional, responsible, and systematic approach to managing risk and apply it unwaveringly to every aspect and activity of the organization.

**Balancing risks at the scene of an incident**    The balancing of risk factors is most critical at the scene of an emergency incident where situational judgment must be exercised by individuals at different levels within the organization--from the incident commander down to the company officer and the individual firefighter who must decide what to do in the face of a critical situation.

*Continued on next page*

## Specific Areas of Concern for the Emergency Service Risk Manager, Continued

**Why risk has to be managed even during non-emergency times**

Many risk factors can be managed in <u>non-emergency</u> times to regulate the level of risk in <u>operational</u> situations.

<u>Examples:</u>
A ladder that is too expensive to test or repair becomes a major liability if it collapses when it is being used at a fire. Self-contained breathing apparatus that has not been thoroughly inspected, tested, and calibrated for over a year might be "good enough" today, but it might not be good enough tomorrow if a medical examiner determines inadequately maintained breathing apparatus caused a firefighter's death.

A poorly trained firefighter might be able to survive dozens of easy situations, yet be unprepared for the first life-threatening emergency he or she faces. A standard operating procedure that is routinely ignored because it is inconvenient for the majority of situations could also be ignored in a situation in which it is critically important. Knowing a building has a wood truss roof might not seem important unless the attic is on fire.

# Chapter Summary

**Key points**      Most persons responsible for risk management in emergency response organizations are not professional risk managers.

Just because something has not happened to one particular fire department in the past does not ensure that it will not happen in the future. The magnitude of a loss *could* be predictable from past experiences, but by the same token, incidents with the most severe consequences are usually the most rare.

Many departments are--to a certain degree--effectively dealing with risk without formal risk management programs.

Examples:
Most departments require firefighters to be trained to a certain level and to wear full protective clothing when responding to a fire, Equipment is inspected and maintained on a scheduled basis and various certifications are required for specified responsibilities.

Such practices are intended to reduce risk. The risk management process incorporates and expands those practices and provides a systematic approach to safety and loss control. The risk management process is intended to provide a comprehensive and detailed system for examining practical and cost effective ways of addressing potential losses.

# Operational Risk Management

# Chapter Overview

**In this Chapter**     This Chapter includes the following topics:

# Risk Management and Emergency Response

**Background**

Operational risk management refers primarily to the risk of death or injury to firefighters and other emergency responders that could result from the performance of their duties. In a broader sense, it applies to other types of accidents and undesirable events that could occur during emergency operations.

Emergency responders knowingly subject themselves to elevated levels of risk in the performance of their duties. Some of those risks are unpredictable and unavoidable. On the other hand, many are well known and can be effectively limited or avoided through the application of operational risk management practices.

**Acceptance of risk**

The acceptance of risk by emergency responders is intimately related to the reasons fire departments and other emergency organizations exist. Firefighters and emergency responders perform essential functions too dangerous for ordinary citizens who are neither properly trained nor adequately equipped, or who might not be physically able to perform such functions. Emergency responders place themselves between the public and a variety of dangers to protect the lives of others. They accept the increased risk to their own lives that is often involved in protecting property.

Risk is an inherent component of the work emergency responders perform. Their ability to work in an elevated risk environment sets emergency responders apart from the general population. In order to survive, emergency responders must effectively manage their exposure to risk by recognizing danger, considering and weighing alternatives, and balancing anticipated benefits with potential consequences. In some cases, doing that leads to the conclusion that a given situation simply does not justify the risk involved in taking action.

*Continued on next page*

# Risk Management and Emergency Response, Continued

**Risk management exists throughout an organization-- TOP MIDDLE BOTTOM**

Risk management occurs at every level of an organization in an emergency operation. It must begin at the top--where the incident commander must determine the appropriate strategy for the incident--and extends down to company officers who must evaluate conditions that define the risk exposure for small groups of workers before they initiate and as they perform their assigned functions. **It extends still further throughout the entire organization--top-middle-bottom--down to the individual firefighter or emergency responder** who must use the same type of judgment to decide on personal actions in many situations.

**Bravery in the face of risk**

Firefighters' reputations are frequently associated with courage and bravery, That perception often suggests that firefighters are willing to accept any risk to their personal safety to perform their duties. Blind acceptance of risk **used to be** virtually unlimited and unquestioned in the fire service. It was not unusual as recently as twenty years ago for firefighters to be exposed to very high levels of risk, with very little concern for their personal safety. Firefighters were expected to follow any order without question and to accept any risk to accomplish the mission. The most respected firefighters were often those with the most obvious disregard for their own safety; those who demonstrated the attitude that the fire must be defeated "at any cost."

Today, we are moving toward a different perception of the relationship between bravery and risk. Without question, we still respect, value, and honor bravery and courage--particularly when a situation involves saving lives. Even so, a contemporary sense of values requires **a very different assessment of appropriate and inappropriate risks.** In many cases, that calls for **limiting the exposure** of personnel to risks that they might be willing to accept for themselves. A fire department's definition of acceptable risk might be more conservative than the level of risk an individual firefighter might willingly accept. In the current value system, higher level officers are often more responsible for limiting risk exposure than for demanding courage from their forces.

*Continued on next page*

**Bravery in the face of risk** (continued)

There are times when only a rescuer's willingness to risk life and limb can save a life. Bravery is still a respected and valued quality for emergency responders. Medals of valor are given to individuals who are willing to risk their own lives to save a total stranger. The general public admires the heroism of firefighters and emergency responders, but **no one expects them to risk their lives where there are no lives to be saved.**

**Today's risk versus protection**

Today's emergency responders have advanced clothing and equipment to protect them, much more capable apparatus and equipment to work with and much better training than previous generations. Even so, they must also face some situations that are much more complicated. Their ability to perform safely and effectively in high risk environments is highly dependent on their ability to recognize the specific dangers that apply to each situation and to work within the limitations of their protective clothing, protective equipment, training, and incident management system (supervision, coordination, and standard operating procedures).

*Continued on next page*

# Risk Management and Emergency Response, Continued

**What is expected**

The citizens and taxpayers who depend on fire departments and emergency service providers for protection have varying perceptions and expectations about the delivery of those services, Because of their willingness to risk their lives to save others, firefighters and emergency responders are often the most respected, admired and valued of all public employees and volunteers, The general public clearly recognizes and values the courage and bravery of all emergency responders, but recognizes that there are reasonable limits as to what can be expected and what is a reasonable level of risk. The expectation is that firefighters will act rationally and professionally--controlling threatening situations without exposing themselves to unnecessary danger.

Fire departments and other emergency response organizations are expected to take every reasonable step to protect their workers from accidents, injuries or disabling occupational diseases. The general public expects fire departments and emergency response organizations to provide their personnel with the training, tools, equipment and support systems that are necessary to perform safely as well as effectively. Taxpayers are generally willing to pay for things emergency responders need, particularly when they are needed to save lives, if those needs are clearly presented. It is not acceptable to the public for fire departments to risk the lives of their members because they are not adequately trained or equipped or because they do not apply appropriate judgment in conducting emergency operations.

*Continued on next page*

**What is expected** (continued)

The public recognizes that many aspects of emergency operations are dangerous and unpredictable. It is almost inevitable that some deaths and injuries will occur due to the nature of the work, but taxpayers do not want to pay for firefighters' injuries that result from taking unnecessary risks. No one expects firefighters to risk their lives where there are no lives to be saved. There is no logical reason for firefighters to risk their lives fighting fires in abandoned structures.

The risk balance for an unoccupied structure might require conducting aggressive offensive fire attack operations, but those operations always involve some degree of risk. Using protective clothing, protective equipment, good training, tactics, supervision and other factors reduce risk, but the risks cannot be eliminated.

If a danger is recognized, it should be avoided. Are any risks acceptable in operations? Only those that are:
- directed toward saving property
- inherent dangers of an unanticipated event
- unknown hazards.

Most buildings and contents are insured, and the owners can protect their property with automatic sprinklers or other systems. Firefighters should not have to risk their lives to save property that can be effectively protected and can be replaced. Those values are incorporated in the basic principles of operational risk management found in *NFPA 1500, Standard on Fire Department Occupational Safety and Health Program.*

**Occupational safety and health laws and emergency operations**

The application of occupational safety and health laws to emergency operations is a relatively recent development. Although worker protection laws were applied to many occupations in earlier days, it is only in the 1980s and 1990s that the same basic principles have been applied to firefighting and other emergency services. The thought that the same principles could be applied to emergency operations as to other occupations and activities is still difficult for many people to accept. As a result, the degree to which occupational health and safety laws, regulations, and voluntary consensus standards are applied is still extremely variable from state to state.

*Continued on next page*

**The fundamental principle of occupational safety and health**

The fundamental principle of occupational safety and health laws is that an employer (the fire department or emergency service organization) is responsible for providing a reasonably safe and healthy workplace for an employee. The application of this concept to emergency operations essentially means an organization must:
- recognize, identify, evaluate the dangers inherent in performing emergency operations
- take reasonable steps to protect employees from those dangers.

Determining what is reasonable in this context generally involves interpreting standards that have been incorporated by reference into regulations such as *NFPA 1500*. The expectation today is that firefighters must be properly trained, supervised, and equipped to function as safely as possible, recognizing the inherent risk factors that are involved in conducting emergency operations.

**Review of operations**

Every incident commander should anticipate that the authority having jurisdiction for occupational safety and health laws will thoroughly review any incidents in which injuries or fatalities occur--using *NFPA 1500* and other applicable standards as benchmarks--to consider if actions taken were reasonable under the circumstances. A fire department should expect that an investigation would seek to determine if its members were provided with every appropriate form of protection, including training and standard operating procedures. There are relatively few cases in which fire departments have been cited for taking inappropriate actions at emergency incidents, however, that has occurred in a few cases and might increase in the future.

The best approach a fire department can take is to regularly review and evaluate its own operations. That way a department can ensure all the components to manage operational risk are in place and that established procedures are consistently applied. The review will address administrative, engineering, and personal protection risk management control methods. If established policies, procedures, training and equipment are appropriate, and if actions taken in conducting operations are reasonable, a detailed external review of operations should not be a cause for anxiety.

# Managing Risk in Emergency Incidents

**Background**

Although risk management is recognized as an inherent and important responsibility of fire departments and other emergency response agencies, it is only in the relatively recent past that the need to actively manage risk during emergency operations has been emphasized. Today we expect the officers who command and supervise emergency operations to function as real-time risk managers. Doing that often involves making critical decisions very quickly and with limited information.

**Risk management is a fundamental responsibility**

Managing risk is a fundamental responsibility at every level of the incident management process.

The specific recognition of risk management is relatively new, but most of the basic principles have always been incorporated in the management of emergency incidents. Fire officers have always been called-upon to make decisions that weigh the risks of a particular course of action against the potential benefits. The major change is the recognition of risk management at the emergency scene as a well defined, value-driven process--not just a matter of personal and situational judgment.

**Develop incident action plan and share it**

An incident commander--who establishes the overall strategic plan for an incident--must identify and evaluate the risks involved in each situation. He or she must make conscious decisions about the acceptable level of exposure to those risks. Supervisors at every level within the incident management structure--who direct and regulate the activities of their subordinates--must be guided by the incident commander's strategic plan, including the risk management considerations incorporated in the strategic plan. All supervisors must also continually evaluate the particular risks present within their assigned areas of responsibility.

Safe operation at any incident will require continuous identification, evaluation, and control of changing conditions. The responsibility **to identify, evaluate, and manage risks** extends to every individual operating at the incident.

*Continued on next page*

68

**NFPA 1500**

*NFPA Standard 1500, Standard on Fire Department Occupational Safety and Health Program* defines the basic principles of operational risk management. The standard establishes specific expectations that the officer in command of an emergency incident will manage the level of risk to firefighters as a fundamental incident management responsibility.

The risk management approach described in *NFPA 1500* should be used as the basis for a fire department's operational risk management policies. A policy statement, supported by a system of standard operating procedures or guidelines, should establish the specific manner in which all members of a department--from the fire chief on down to the newest member--expect to operate at emergency incidents. The principles have been widely adopted by many fire departments as the standard operational approach to risk management.

The section that specifically refers to operational risk management was introduced in the 1992 edition of *NFPA 1500.* It is reproduced on the following pages, Note that the text in plain type is the mandatory requirement. The portions in italics are part of the *NFPA 1500 Appendix,* which is explanatory and supplemental.

*Continued on next page*

---

### From NFPA 1500 - 1992 Edition

6-2 Risk Management

6-2.1 The incident commander shall integrate risk management into the regular functions of incident command.

*A-6-2.1 The incident commander has an ultimate responsibility for the safety of all fire department members operating at an incident and for any and all other persons whose safety is affected by fire department operations. Risk management provides a basis for*
*(a) Standard evaluation of the situation*
*(b) Strategic decision making*
*(c) Tactical planning*
*(d) Plan evaluation and revision*
*(e) Operational command and control*

6-2.1.1 The concept of risk management shall be used on the basis of the following principles:
(a) Activities that present a significant risk to the safety of members shall be limited to situations where there is a potential to save endangered lives.
(b) Activities that are routinely employed to protect property shall be recognized as inherent risks to the safety of members, and actions shall be taken to reduce or avoid these risks.
(c) No risk to the safety of members shall be acceptable when there is no possibility to save lives or property.

*A-6-2.1.1 The risk to fire department members is the most important factor considered by the incident commander in determining the strategy that will be employed in each situation. The management of risk levels involves all of the following factors:*
*(a) Routine evaluation of risk in all situations*
*(b) Well-defined strategic options*
*(c) Standard operating procedures*
*(d) Effective training*
*(e) Full protective clothing and equipment*
*(f) Effective incident management and communications*
*(g) Safety procedures and safety officers*
*(h) Back-up crews for rapid intervention*
*(I) Adequate resources*
*(j) Rest and rehabilitation*
*(k) Regular evaluation of conditions*
*(l) Pessimistic evaluation of changing conditions*
*(m) Experience based on previous incidents and critiques*

---

*Continued on next page*

# Managing Risk in Emergency Incidents, Continued

---

**From NFPA 1500 - 1992 Edition**

6-2.1.2 The incident commander shall evaluate the risks to members with respect to the purpose and potential results of their actions in each situation. In situations where the risk to fire department members is excessive, as defined by 6-2.1.1 of this section, activities shall be limited to defensive operations.

*A-6-2.1.2 The acceptable level of risk is directly related to the potential to save lives or property. Where there is no potential to save lives, the risk to fire department members must be evaluated in proportion to the ability to save property of value. When there is no ability to save lives or property, there is no justification to expose fire department members to any avoidable risk, and defensive operations are the appropriate strategy.*

6-2.2 Risk management principles shall be routinely employed by supervisory personnel at all levels of the incident management system to define the limits of acceptable and unacceptable positions and functions for all members at the incident scene.

6-2.3 At significant incidents and special operations incidents, the incident commander shall assign qualified personnel with the specific authority and responsibility to evaluate hazards and provide direction with respect to the safety of operations.

*A-6-2.3 A safety sector should be established at all major incidents and at any high risk incidents. The safety sector would normally be assigned to operate under the fire department safety officer or an assigned officer with this responsibility. If the designated safety officer is not available and the need for a safety officer is evident, the incident commander should assign one or more individuals to assume this responsibility on a temporary basis. Depending on the specific situation, this assignment could require one or more members. All members should be familiar with the basic duties and responsibilities of a safety sector:*

---

*Continued on next page*

---

**Systematic approach required--all members must be involved**

The individual in command of an incident is specifically responsible for managing risk at the incident, however, one person cannot be expected to apply these principles to an incident if the organization has not integrated a standard approach to risk management into its standard operating procedures and its organizational culture. To be effective, risk management principles and policies must be integrated into the entire operational approach of the fire department or emergency response organization. They must be incorporated within the duties and responsibilities of every officer, supervisor and member.

Most fire officers seem to feel that the risk management principles stated in *NFPA 1500* are generally in agreement with their personal beliefs. Still, many fire officers initially expressed discomfort with the presence of a written policy, as opposed to leaving a wide span of discretion to the individual incident commander's judgment. Their discomfort might have been a reaction to changing from a system in which all command decisions were considered prerogatives of an incident commander to a system in which an incident commander can be held accountable for applying written policies.

---

**Risk assessment**

The most important and difficult concept in the operational risk management process is the actual determination of the types and levels of risk that are present in each situation and the degree of risk that is acceptable for the personnel who are operating at that incident. There is always some degree of risk involved in conducting emergency operations. The incident commander has to determine the limits of risk that are acceptable for each situation and direct operations to ensure that those limits are not exceeded.

---

*Continued on next page*

**Three guidelines define acceptable levels**

There are three simple guidelines that define acceptable levels of risk:

1. Activities that present a significant risk to the safety of members shall be limited to situations in which there is a potential to save endangered lives,
2. Activities routinely employed to protect property shall be recognized as inherent risks to the safety of members, and actions shall be taken to reduce or avoid those risks.
3. No risk to the safety of members shall be acceptable when there is no possibility to save lives or property.

Those three statements place major responsibility on an incident commander to first, identify and evaluate the risks that are present in each situation and then exercise good judgment to determine when the level of risk is excessive. Without good judgment, risk management policies are nothing more than words on paper.

**Risk management drives the strategic plan**

The strategic plan for a fire defines **where, when, and how** firefighters will seek to control that fire. An incident commander's most fundamental responsibility is to establish and implement a strategic plan to conduct an emergency operation. He or she must always weigh the exposure to danger against the anticipated results of a strategic plan. **Risk management is one essential consideration in developing a strategic plan.**

An incident commander's choice of operational mode--**offensive or defensive**--defines the "rules" that apply to everyone involved in that incident. In offensive situations, firefighters enter burning structures and attempt to control a fire where it is burning. They expose themselves to all the risks present in that environment. In the defensive mode firefighters avoid many of those risks by staying out of the most dangerous areas and conducting operations that limit the spread of a fire to an area that can be defended without exposing participants to unnecessary risks.

*Note:* A parallel exists with every emergency operation in which incident commanders must decide if the potential benefits justify exposing personnel to the risks that are present.

*Continued on next page*

**Initial risk assessment**

The determination of acceptable risk and the choice of operational mode begins with the first arriving officer, who must assume command of the incident and make an initial size-up. Even if the decision applies to only the first arriving company, basic risk management policies must be applied to determine appropriate actions--including the initial choice between offensive and defensive operations.

At a fire incident, the first arriving company officer often has to decide if and when it is appropriate to enter a burning structure to conduct offensive operations. The choice depends on risk factors that are identified with respect to the size, location and stage of the fire and the capabilities of the fire suppression force.

Before an interior attack can be initiated, the officer must be assured that firefighters and equipment available on the scene have the capability to conduct a safe and effective interior operation. The initial risk assessment must consider the possibility of saving lives, because the acceptable level of risk to save a life is higher than the acceptable risk to protect property. It must also consider the fire conditions and the risks they present to firefighters. If the first unit does not deliver enough personnel to conduct a reasonably safe interior attack operation, the interior attack might have to be delayed until additional personnel arrive. If the situation is too dangerous for safe entry, the plan should be limited to exterior operations.

*Note:* A similar determination must be made for many other types of emergency incidents. For example, in a trench rescue situation, the decision would be related to whether or not it is safe to have the first-arriving personnel begin to dig to rescue a trapped victim. If the risk to rescuers is excessive, a rescue attempt must wait until a specially trained trench rescue team arrives with the appropriate equipment and shoring. The decision depends on where and how the victim is trapped and the risk of further collapse that could turn the rescuers into victims. Similarly, in a confined space rescue incident, the first arriving officer must determine if personnel at the scene have the skills and equipment that are needed to initiate a rescue operation. If not, the rescue must wait until a specialized team responds.

*Continued on next page*

**Risk factors known and assumed**

It is often impossible to obtain and verify all the pertinent information before making important strategic decisions. The decisions that guide initial actions must often be made with a limited amount of information, making allowances for factors that are not known. A rapid size-up seldom allows for a full evaluation of all the risk factors. An incident commander must consciously differentiate between factors that are "known" and those that are based assumptions, experience and standard approaches and must then place a priority on either confirming these assumptions or revising them as soon as factual information can be obtained.

**Ongoing incident commander responsibilities**

An incident commander is responsible for determining the operational mode-- offensive or defensive--for the entire duration of the incident. After making an initial strategic determination, an incident commander should define tactical objectives and assign resources to perform specific functions. He or she must also begin to establish a command structure to effectively supervise the operation. Whenever command of an incident is transferred, the responsibility for strategic risk management is transferred too.

**Responsibilities at all levels**

Officers and supervisors assigned at each level of the incident management organization must apply risk management policies within their areas of responsibility--following the strategic plan and the associated "rules" established by the incident commander. Keeping the incident commander informed is one of their major responsibilities. That's especially true if changing conditions indicate the need to change the strategic plan.

The single most important reason to establish an effective incident management structure is to ensure that operations are conducted safely. Every individual in the incident management structure is responsible for monitoring and evaluating risks and for keeping the incident commander informed of any factor that could be a reason to reconsider the risk management balance.

A risk assessment should be reprocessed with every observation and progress report. Every bit of new information that comes to the incident commander should be considered to see if strategy or tactics should be changed.

*Continued on next page*

**Roles and responsibilities**

Every individual in the emergency response system has a role in operational risk management, as the table that follows indicates.

| Role | Responsibilities |
|---|---|
| Incident Commander | Is expected to make strategic decisions based on risk acceptance or avoidance. Decisions are incorporated in the selection of offensive or defensive operations, a major component of the strategic plan for the incident. |
| Sector and company officers | Are expected to supervise tactical operations based on risk acceptance/avoidance in the situations they encounter within the areas they supervise. Their determinations must be consistent with the direction provided by the incident commander. They must keep the incident commander informed of any situations they encounter that may have an impact on the strategic plan. |
| Individual firefighters/ EMS personnel | Might be called on to make personal decisions about risk acceptance/avoidance when no officer is present, which must also be consistent with the strategic plan and with departmental policies. They also need to keep their officers informed of any significant information. |
| Incident Safety Officer On-Scene Risk Manager | Is the risk management consultant for emergency operations. The incident safety officer is an advisor to the incident commander and should provide an overview of the situation specifically directed toward identifying and evaluating safety concerns. |

*Continued on next page*

**Gathering additional information**

The initial strategic plan is always subject to revision as the amount of confirmed information increases and the degree of uncertainty decreases. An incident commander must constantly seek out information to replace assumptions or perceptions with verified facts. Officers assigned to supervise different areas or functions are expected to provide regular progress reports to an incident commander. In addition, an incident commander must ask questions and actively seek information that is not always provided without prompting.

An incident commander should not hesitate to assign individuals to reconnaissance missions--to observe conditions firsthand, or seek out information and report back to command post. An incident commander should always have this capability, even if it requires calling for an additional company or command officer to perform this important function. At major incidents, reconnaissance can often be assigned to staff personnel who report to the command post, especially to individuals known to be particularly capable at gathering and managing information.

**The role of the incident safety officer**

There should be a standard system for assigning an incident safety officer at working incidents, preferably by dispatching someone (or more than one person) with this specific responsibility. Whenever possible, the incident safety officer should begin by making a 360 degree survey of the incident scene to evaluate the overall situation--looking for:
- problems
- hazards
- inconsistent observations
- any other factor that could indicate a safety concern.

After making an initial survey, the incident safety officer should report to the command post to discuss his or her observations and safety concerns with the incident commander.

*Continued on next page*

**Balancing perspectives**

The incident safety officer does not relieve an incident commander of the responsibility for managing risk at an incident. By the same token, an incident commander should be able to rely on the incident safety officer to provide a balancing perspective on the situation.

---

**An Incident Commander should look at a situation as: "How to get the job done and operate safely."**

**The Incident Safety Officer should look at the situation as: "How to operate safely and still get the job done."**

---

In most cases, the two perspectives should lead an incident commander and the incident safety officer to reach the same conclusions. If they do not, the incident commander must recognize the possibility of a problem and rethink the risk analysis.

# Application of Risk Management Policies

**Background**

An incident commander has to successfully balance two sets of factors:
- those that determine the nature and extent of operations
- those that address the safe conduct of operations

This section explores some of the complexities involved in trying to <u>consistently</u> apply risk management policies to determining the nature and extent of operations.

The operational risk management guidelines defined earlier establish a simple framework that should serve as the basis for operational risk management decisions. Many situations are complicated. It is often difficult to interpret how a policy--stated in general terms--relates to a specific situation. An incident commander is expected to:
- use judgment and experience
- make reasonable decisions in the application of policy guidelines.

An incident commander has a wide span of discretionary authority for making risk management decisions. A strategic plan must not needlessly place the lives of firefighters or emergency responders in danger, but it should not be so over-cautious that it allows a fire to destroy property that could be saved--or keeps other valuable functions from being performed. The ultimate test of a risk management decision is whether or not a reasonable, well-informed person would find the decision appropriate under the circumstances.

**Experience and judgment required**

The circumstances responders actually encounter in emergency incidents are often very complex. That fact can make it extremely difficult to apply the policy guidelines to each particular situation. The application of policy guidelines requires specific skills at obtaining, considering, and applying information. A combination of experience and judgment are essential to perform the actual evaluation of the risk factors.

*Continued on next page*

# Application of Risk Management Policies, Continued

**Unoccupied, vacant, and abandoned buildings**

As the table below indicates, the distinguishing characteristics of unoccupied, vacant, and abandoned structures influence incident commanders' decisions. Fires in vacant, unoccupied or abandoned buildings are a problem to many fire departments. The fundamental risk management guidelines state that the lives of firefighters should not be risked where there is no possibility of saving lives or property. That would restrict entry into many vacant or abandoned structures to conduct interior operations--on the basis that there are no occupants and there is nothing of value to be saved.

| Describing a building or structure as... | Implies... |
|---|---|
| Unoccupied | That it has contents and is suitable for occupancy, however, no one is present at the time of an incident. |
| Vacant | That the interior space is not currently in use and presumably has neither contents of value or occupants. However, the structure itself might be considered valuable property. |
| Abandoned | That it has no value that justifies the level of risk associated with an interior offensive operation. |

The first priority in a fire suppression operation is rescue. Many fire departments operate with the assumption that there could be occupants in any building. As long as it is reasonably safe to enter, they attempt to conduct a search of every occupancy. If the structure (or part of it) is fully involved in flames, there is no reasonable possibility that anyone could be alive to rescue and no reason to conduct an interior search.

Unoccupied and vacant structures could justify operations that involve a limited amount of risk. It could also be appropriate to enter an abandoned building to conduct a rapid search if there is reason to believe that it might be occupied by transients or other "unofficial" occupants. The deciding factors to be weighed include:
• the extent of the fire
• the structural condition of the structure, and
• whether it is truly abandoned or simply vacant.

*Continued on next page*

# Application of Risk Management Policies, Continued

**Judging the risks of abandoned structures**

Firefighters arriving at a small fire in an abandoned building might have to consider the risks they could encounter in an interior attack versus the consequences of not attacking the fire. There might not be any justification to risk the lives of firefighters to save a worthless structure, but it could be possible to justify making entry to extinguish an incipient stage fire or a fire that is contained within a small area. If the fire can be safely controlled, an interior attack might involve less risk than staying outside and letting the fire grow.

This type of attack is justifiable only when not attacking and controlling the small fire would expose firefighters to a greater risk. If entry is made in this type of situation, it must be very cautious and closely supervised.

**Risk to the community**

The risk evaluation might have to consider more than the structure that is on fire. A fire that grows to fully involve a large abandoned building could create a significant risk of fire spread to occupied exposures or to the surrounding area. The incident commander might have to weigh the risks of an offensive attack against the risk of fire spread to the exposures.

If the exposures can be protected without exposing the surrounding community to excessive risk, the incident commander can evaluate the risks to firefighters in relation to the building that is on fire. If the resulting fire cannot be contained to the abandoned building, the risk evaluation should also consider the potential consequences of a fully involved structure.

*Continued on next page*

**Unusual situational risks**

There are many additional examples of situations in which good judgment is clearly required, considering the health and safety of firefighters in relation to the potential consequences of different strategies. The potential consequences often have to be considered in determining the level of risk that is acceptable.

Example:
A building could be known to contain hazardous materials that would cause a massive contamination problem if water runoff carried them into a nearby river. That possibility could make it unacceptable to fight a fire, causing the fire department to limit control actions to protecting exposures.

Example:
In another situation--should a fire be allowed to burn-- the release of toxic products of combustion could result in the greatest risk. In some cases, the risk of exposing firefighters to hazardous materials could make it unacceptable to enter. In other cases, entering and controlling the fire before it reaches the hazardous materials could be the best plan to protect the firefighters and the community from a greater risk.

*Continued on next page*

## Application of Risk Management Policies, Continued

**Unusual structural situations**

Firefighters should never assume a situation is "routine." They should always be looking out for situations that present unusual risks. Well developed sources of information can uncover hundreds of factors that might draw immediate attention to the possibility of high risk. When any of those factors are seen or reported, an incident commander should automatically:
• reprocess risk management decisions already made
• reevaluate the strategy adopted for the incident.

In <u>Fire Command</u> (National Fire Protection Association, 1985), Chief Alan Brunacini of the Phoenix, Arizona Fire Department points to three components that comprise incident information management:
• visual   reconnaissance
• pre-fire   planning
• familiarity  with  the  location  or  situation

<u>Examples</u>:
Buildings with grade level entrances on different sides that lead to different interior floor levels. That could cause confusion about the level where different crews are operating--above, below or on the same level as the fire.

Buildings that have been remodeled, where interior supports may have been removed or replaced with inadequate construction.

Buildings with extra roofs, floor coverings, or ceilings that conceal the original construction. Such spaces make it impossible to evaluate the structural characteristics, and might allow a fire to grow and extend.

Buildings with engineered lightweight structural systems, such as lightweight wood trusses that are susceptible to sudden and early collapse.

Buildings with unexpected interior shafts or stairways that could allow a fire to extend  to  an  upper  level.

Buildings with complicated interior arrangements and long travel distances to entry  and  exit  points.

Interconnections between buildings that cannot be easily interpreted from the exterior arrangement.

*Continued  on  next  page*

**Evaluating information from different sources**

While conducting operations an incident commander should carefully evaluate information and reports that come from sector officers or other individuals and consider their consistency. Communication starts with knowing how to listen carefully and critically. If information does not appear -to make sense, an incident commander should question and verify it. An incident commander should always take a pessimistic view of conflicting or uncertain information. It is all right to hope for the best, but plans and actions should always anticipate the worst predictable outcome.

Examples:

Reports from different sources that describe inconsistent fire conditions. One observer might report "fire under control" while another reports heavy smoke or fire conditions. The discrepancy suggests they are looking at different areas or that one does not know about fire conditions that are evident to the other. Crews operating in the area where the fire appears to be under control might be in serious danger if they do not know where the fire is actually or still burning.

Evidence of a significant interior fire that cannot be located should sound a warning to the incident commander. Crews working in a smoke filled building might be unable to find the fire at the same time the continuing or increasing presence of heavy smoke suggests that a significant fire is burning somewhere inside the structure. The risk of a sudden outbreak of fire or a structural collapse increases with time spent on the scene.

An incident commander and other officers should look for unusual colors or movement of smoke. Smoke movement often provides good information about the size and extent of a fire. A distinct thermal column and rapid smoke movement would suggest the presence of a hot fire, possibly in a concealed space.

*Continued on next page*

# Application of Risk Management Policies,

**The time factor**

One of the major challenges in operational risk management is the time factor. The most critical decisions often have to be made very quickly and under great pressure. An incident commander must be an efficient decision maker, able to quickly:
- recognize dangerous situations
- evaluate information
- make decisions.

**Some emergency responders tend to be more inclined to take action than to stop and compare the risks that could be involved in alternative approaches.** Individuals who have to make risk management decisions must be able to gather and process information efficiently and to think clearly and quickly in stressful situations. In some cases, they must slow down action-oriented individuals to provide time to evaluate the situation.

Time tends to work against a decision maker in high stress situations. The time that is available to make decisions passes very quickly, while the time it takes for information to be gathered and reported seems to take forever. It is easy to lose track of time in a stressful situation. A system that reminds the incident commander when each ten or twenty minute period has elapsed is a valuable addition to an incident management system.

In many cases, particularly in fires where the structure is deteriorating as long as the fire continues to bum, the risk factors will also change with time. If a fire is not controlled, interior crews must be withdrawn and regrouped in safe areas before a structural collapse occurs. Waiting to see what happens may prove to be a fatal error.

**Rules of thumb**

There are some "time factor rules of thumb" for structural collapse:
- an ordinary construction building is susceptible to structural collapse after twenty minutes of fire involvement.
- a lightweight wood truss structure may collapse within ten minutes,
- a fire in a fire-resistive building may withstand full involvement of a fire area until all of the available fuel is consumed.

*Continued on next page*

# Application of Risk Management Policies, Continued

**Unacceptable risk**

There are many situations that require an incident commander to exercise judgment to determine the acceptable level of risk. There are other situations in which--under almost any conceivable circumstances--the level of risk is clearly unacceptable. When an incident commander encounters such a situation, he or she must be prepared to direct a course of action that avoids the danger.

Examples:
Fighting fires that could involve explosives.

Continuing to conduct interior operations in a structure that exhibits signs of imminent structural collapse, such as might result from structural weakness of temporary repairs once made to correct structural defects. (Any individual who spots evidence of such a condition should immediately see to it that the incident commander is made aware of it.)

A fire department officer should:
- have a good knowledge of situations that are outside the limits of acceptable risk exposure
- know how to quickly recognize and react to them.

Many risk-laden situations can be identified through inspection activities or pre-fire planning visits. When situations are found that would create an unacceptable risk, they should be documented in a manner that:
- supports training and hazard awareness
- provides critical information for an incident commander should an incident occur at the location.

*Continued on next page*

86

**Focus**

There are often many distractions at the scene of an incident that could keep an incident commander from focusing on, and processing important risk evaluation factors. That possibility reinforces the importance of making risk assessment a standard part of the process of commanding incidents. If an incident commander always thinks about risk factors, in every situation, risk evaluation will become a priority among the distractions of a complex situation.

Thinking ahead of the incident will also help an incident commander make difficult decisions under time pressures, Many situations and circumstances are predictable--whether it is an anticipated fire in a particular building or a predictable situation that could occur in a variety of locations and circumstances. If potential situations can be predicted and the risk factors can be thoroughly evaluated in advance--when there are no time pressures--an incident commander only has to recognize that situation when it occurs and implement the planned strategy.

Example:

If an imminent risk of structural collapse is identified, the incident commander should be prepared to immediately withdraw all companies from interior or exposed positions, call for a roll-call accounting of all personnel, then re-group the operating companies for defensive operations.

# Conducting Operations Safely

**Background**

The previous section dealt with the determination of the *nature and extent of operations,* based on reducing the potential for risk to firefighters and emergency responders at a reasonable and acceptable level in a wide variety of situations. In this section, we will give equal consideration to the *safe conduct of operations*--operations that limit risk to an acceptable level.

Avoidance of operational risks must begin long before an incident occurs. For example, **many operational risks could be avoided if fire prevention activities were effectively implemented.** When prevention fails, fire departments respond and must face operational risks. To fully avoid operational risk, a fire department would have to take such a cautious approach that it would accomplish very little in the sense of saving lives or property. In many cases, the best way to minimize overall risk is to take decisive action--using all the force necessary to effectively control a problem while it still can be managed.

Emergency operations must always be conducted as safely as possible. **Even during potential rescue situations, in which the maximum exposure to risk is permissible, standard safety rules are not suspended.** Whether the acceptable level of risk is high or low, the operation must support safe operations with every available skill and technological resource.

**Definition of acceptable risk**

An acceptable level of risk means that--after taking every reasonable precaution--there still exists a recognized possibility that firefighters could be killed or injured. The threshold of acceptable risk is set higher when there is a possibility of saving lives, not just property. This implies that it is acceptable to risk the lives of firefighters to save property, however no property is worth risking the life of a firefighter. The reality is that we can never eliminate the possibility that something could happen that would result in an injury or fatality but the **possibility** of injury or death must never be allowed to become a significant **probability.**

*Continued on next page*

# Conducting Operations Safely, Continued

**Accepting inherent risks**

The risks that are inherent in firefighting and other types of emergency operations are considered acceptable, yet they are significant. After all, it is very possible to be seriously--perhaps fatally--injured as the result of an inherent risk. Why then are inherent risks generally considered acceptable? The risks are considered acceptable only because firefighters and emergency responders are:

- trained
- equipped
- prepared
- organized to conduct operations with a reasonable degree of safety in dangerous environments. Their preparedness makes the inherent risks acceptable.

**Two modes of operation**

Operational firefighting strategy can be based on two distinctly different modes of operation: offensive and defensive. The table below describes both modes.

| Mode of Operation | Description |
| --- | --- |
| Offensive | Firefighters are in close contact with a fire--inside a building, exposed to all the potential dangers that exist in that environment. They are dependent on their protective clothing and equipment for protection. |
| Defensive | Firefighters should be outside the area of direct exposure. Still, many things could happen to injure firefighters. For example, the incident commander must ensure that all personnel stay outside an area identified as dangerous, including outside the range of falling walls. |

*Continued on next page*

**Why incident commanders might change strategy**

Often, the story of an incident unfolds like this: The initial action occurs in the offensive mode because occupants are believed to be inside and in need of rescue. Then conditions change. They reach a point at which there is no further possibility that lives can be saved. All occupants may have been removed and accounted-for, or the fire may have grown beyond a point where anyone could survive. When that happens, the incident commander has to reevaluate the original determination of acceptable risk, now based on the possibility of saving property. The overall situation might no longer support continued interior offensive operations. If the incident commander determines that the potential benefits no longer justify the risks, he or she must change the strategy from interior offensive operations to exterior defensive operations.

**When to change strategy**

In any event, an incident commander must change the operational strategy whenever the level of risk to operating personnel deemed to be acceptable is exceeded--either because the situation has changed from a life *saving* to a *property saving* operation or because an interior offensive fire attack is not controlling the fire. If that happens, firefighters must be removed from the interior before they are trapped by flames or buried by a structural collapse.

An incident commander must always anticipate the likelihood of deteriorating conditions and be prepared to change his or her strategy on extremely short notice.

The incident commander should be outside at a command post where it is possible to evaluate the "big picture" and direct operations in an environment that is conducive to managing information and communications. The incident commander often has to depend on other individuals to provide information about changing conditions that are not visible from the command post, particularly interior conditions. At the same time, interior crews have to depend on the incident commander's ability to evaluate conditions, because their ability to evaluate a situation may be very limited. (It is difficult to evaluate conditions effectively while wearing breathing apparatus and operating a hoseline inside a smoke-filled building.) The incident commander has to depend on others to report information, particularly risk-related information and the interior crews have to trust the incident commander's judgment to decide when it is time to retreat.

*Continued on next page*

**Implementing a change in strategy**

When the strategy is changed from offensive to defensive, participants are removed from the danger area to operate from positions where there is the least possible risk of accident or injury. To implement the change, an incident commander should:

- order all interior attack crews to immediately abandon their efforts and evacuate to safe position. (This can be accomplished by a distinctive radio notification and by other methods, such as sounding apparatus air horns.)
- conduct a positive accountability check to ensure that all personnel have evacuated and removed themselves to safe positions.

Interior crews must be prepared to react to a change in strategy without hesitation or delay. An incident commander should initiate exterior attack operations only after verifying the safety of all personnel through the personnel accountability system.

**Standard operating procedures**

Standard operating procedures or guidelines establish the basic framework for conducting emergency operations safely and effectively. Some procedures are specifically directed toward safety; others support safe operations by establishing a system of predictable and consistent operations. Coordination, consistency, and standard approaches are all important and valuable components of safe emergency operations. One key component of operational risk management is to ensure that operations are always conducted in a standard manner that incorporates a full range of safety considerations.

*Continued on next page*

# Conducting Operations Safely, Continued

**The need for a consistent approach to safety**

Consistency in operational risk management begins with:
- establishing safety procedures
- fully documenting them
- train all members of the organization in the application of those procedures.
- ensure that required procedures are consistently implemented at every incident.

A consistent approach to safety requires that all the applicable standard operating procedures are actually followed. The only thing worse than having no safety procedures might be to have procedures that are not consistently applied and enforced.

---

**Consistent Approach To Safety**

DO THE RIGHT THINGS

DO THEM THE RIGHT WAY

ENSURE THEY ARE BEING DONE THE RIGHT WAY

---

*Continued on next page*

**Components of a basic operational system**

Important safety-related components and considerations that should be established within a fire department's basic operational system include all the components listed below.

---

### Basic Operational System
### Components and Considerations

### Information from pre-fire planning
### Communications
### An accountability system
### Rapid intervention teams
### Rest and rehabilitation

---

We will examine each component.

---

**Information from pre-fire planning**

Incident commanders are generally pressured by time and must rely heavily on gaining information rapidly visually and by reconnaissance. Commanders usually acquire visual information by personally surveying the scene. Reconnaissance information is usually gathered by others who are assigned to this task by the incident commander.

An incident commander can also obtain valuable information from formal pre-fire planning from other agencies, or from other activities that provide useful information. Having access to previously generated sets of facts about a structure or a given situation can:

- provide an incident commander information that is more complete than would be available through other methods
- allow the incident commander to reallocate personnel who might have been assigned to reconnaissance and obtaining information on the scene
- save valuable time that would be needed to gather information.

Pre-fire planning and the management of information are critical components of the incident risk management system.

---

*Continued on next page*

# Conducting Operations Safely, Continued

**Communications**     To direct an emergency operation, an incident commander must have an effective communications system. All interior crews should have portable radios to maintain contact with the incident commander or their sector supervisor. A reliable two-way radio link between an incident commander and each operating sector officer, group, company, attack team or other organizational unit provides the means for:
- operating personnel to call for assistance if they need to be rescued
- the incident commander to direct personnel to evacuate a building or take other actions when he or she recognizes an imminent hazard is recognized or reported.
- all personnel to be aware of any changes in the incident action plan.

A standard protocol should be established for transmitting emergency messages to ensure that they receive priority over all other radio traffic. The radio channel used for tactical communications at the incident scene should be reserved for that purpose to ensure that unrelated radio traffic does not block out critical messages.

The tactical channel should be used by all units working at the incident and a designated individual at the command post must constantly monitor the channel for messages related to safety-related situations or emergencies, Where multiple radio channels are used, there should be one dedicated channel to be used by any unit that needs emergency assistance. That channel must be constantly monitored at the command post.

*Continued on next page*

**A personnel accountability system**

A personnel accountability system should be routinely employed at emergency incidents to keep track of the location, assignment and welfare of all personnel operating in hazardous or potentially hazardous areas. The system should keep track of individuals assigned to each company or team working at the incident scene. It should be used to verify the status of each company or team at regular intervals. An immediate priority must be placed on locating anyone who is not confirmed to be safe when an accountability check is made. A full accountability check should be made when designated events occur, such as:

- an evacuation of interior crews
- a switch to defensive strategy
- a report of personnel in trouble
- the reaching of a benchmark

The ability to use an effective accountability system depends absolutely on:

- an effective incident organization
- a realistic span of control
- capable supervisors at every level
- recognized responsibilities at all levels
- all players buying into the process

**Rapid Intervention Teams**

The assignment of one or more Rapid Intervention Teams at working incidents provides the ability to immediately initiate a rescue effort to locate, rescue or assist any firefighters (or anyone else) who are in trouble at the scene of an incident. Rapid Intervention Team members should be standing-by wearing their protective clothing with self-contained breathing apparatus ready for immediate use. They should have forcible entry tools, rescue rope, and any other equipment that could be needed quickly. At hazardous materials incidents or other situations where special protective equipment is required, the Rapid Intervention Team should be ready with the same level of protective clothing and equipment as the entry team requires.

*Continued on next page*

# Conducting Operations Safely, Continued

**Rest and rehabilitation**

Crews who are fatigued are much more susceptible to accidents and injuries than fresh, well rested crews. To avoid fatigue and exhaustion among crew members, the standard operational approach should provide:
- monitored periods of rest
- rehabilitation
- fluids at regular intervals
- medical evaluation

**Overhaul and salvage**

Many fireground injuries occur after the fire is under control, during salvage, overhaul, investigation, and decommitment phases. That is when life safety and property protection are no longer urgent priorities. That is when the safety of firefighters should remain the primary concern. It is often advisable to withdraw crews and assign sector officers or the incident safety officer to make a complete survey of the damaged area before assigning personnel to complete the remaining tasks. This survey can identify hazards that must be avoided and determine where it is safe to operate. It also provides time to set up portable lighting and ventilation fans to eliminate hazards before assigning any crews to work in these areas.

*Continued on next page*

# Conducting Operations Safely,

**Limiting risk by carrying out assigned responsibilities**

Fire suppression operations should be conducted by well-trained, well-equipped firefighters, operating under an effective incident management system and working in teams under the supervision of capable officers to perform <u>assigned</u> tasks that have been coordinated within an incident action plan.

The incident commander establishes the overall strategy that defines the acceptable risk level for an incident. The responsibility for managing risk extends throughout the incident management structure. It includes every level--down to the individual firefighters who must be able to recognize hazards and avoid unnecessary risks. The incident commander must effectively communicate the strategic plan to every part of the organization and ensure that everyone understands his or her assignment.

Whether or not they agree with the incident commander's assessment of the situation, subordinates are expected to follow the directions that come down the chain of command. They have to feed information back up the chain of command to keep their incident commander aware of what is happening.

There is no absolute measure or definition of an acceptable level of risk exposure for individual firefighters in the performance of their duties. Some individuals or crews might be willing to accept a higher level of risk than the organization or the incident commander are willing to authorize. From their limited perspective, they might be unaware of critical information, or they might simply be willing to expose themselves to a higher level of risk.

A fire department establishes the context in which risk acceptance/avoidance decisions are made. Risk management policies and approaches should be part of the organizational culture, defined by policy, supported by training and applied with experience, and judgment. **Operating outside the established system should never be accepted.**

A consistent approach and application of safety procedures should establish a basic level of operational safety for the organization. At each incident, the incident commander adds situational judgment that defines the level of risk exposure for that incident. The operational discipline of the organization must require everyone to follow orders--even if they feel their actions are being unnecessarily   restricted.

*Continued  on  next  page*

**Personal protection**

The ability of fire departments and emergency response organizations to operate safely is constantly improving as improvements are made in personal protective clothing and equipment. These advances have actually reduced the risk of injury or death during operations that were already considered "reasonably safe" in many situations. The margin of safety has been improved for users who are performing the same operations.

> **Members must take time to ensure they are wearing appropriate personal protective clothing and equipment. Check each other. No exceptions.**

The improvements in personal protection have also expanded the range of conditions where firefighters can operate without being injured. In the past, firefighters' level of exposure was often regulated by the limited protection provided by their protective clothing. Working with exposed ears, poorly insulated gloves, and wearing coats and 3/4 length boots that provided limited insulation, firefighters were able to sense the heat. They were often stopped from further penetration by the threshold of pain. First and second degree bums to ears, knees, and wrists often provided evidence that firefighters had advanced as far as they could into a superheated atmosphere.

*Continued on next page*

| | |
|---|---|
| **Protection is limited** | Firefighters should wear full protective clothing ensembles that meet current standards. Breathing apparatus and PASS should be checked regularly and function-tested at regular intervals to ensure they will perform reliably. The fit of face pieces should be tested regularly. Firefighters should be trained to achieve a high level of confidence and familiarity with their breathing apparatus. Major advances--based on extensive monitoring of the environment to which firefighters were exposed--were made in protective clothing and equipment during the 1980s. Levels of exposure to different temperatures were evaluated to establish performance standards for protective clothing--including relative exposure to radiant, convective and conducted heat. The new generation of protective clothing, which provides full body protection, is designed to prevent bums in the environment that was measured and to protect a firefighter who is caught in a flashover for ten to twenty seconds. The level of risk to the firefighter is significantly reduced in relation to the atmosphere for which it was designed. |
| | Personal protective clothing and equipment provides limited protection from the heat of a fire. It is very difficult to control the direct exposure of firefighters who are engaged in interior fire attack operations. In many cases, only the individual members who are actually making the attack are in a position to evaluate the interior fire conditions. |

*Continued on next page*

**Personal protection benefits and concerns**

Improved protection might also allow users to increase their penetration into a fire atmosphere--knowingly or unknowingly. Although many injuries have been prevented or reduced in severity with the protection that the new clothing provides, the risk balance shifts in the wrong direction if the extra protection causes firefighters to be exposed to a more dangerous situation.

How could that happen? Firefighters might be able get in deeper and control fires where they could not previously penetrate, but they might also be <u>unable</u> to sense the danger of a hotter atmosphere or rapidly increasing temperatures. In this situation, they might unwittingly expose themselves to an increased risk of severe injury.

An incident commander at an exterior command post cannot manage this type of risk. If conditions appear to be too dangerous from an incident commander's vantage point, he or she must withdraw firefighters from the interior. The operation must switch from offensive to defensive operations.

It is virtually impossible for an incident commander outside a structure to fully evaluate conditions inside. In most cases, the risk of penetration into the fire atmosphere can be monitored and evaluated only at the place it occurs. An incident commander has to depend on company officers, sector officers and the incident safety officer to help him or her:
• observe and evaluate conditions
• monitor the level of exposure
• control the actions of the attack crews.

The advances that have been made in protective clothing and equipment must be matched with closer supervision, coordination, and communications. Training, standard operating procedures and experience are also important factors in controlling the exposure to excessive risk and conducting safe operations.

# Preparation for Incident Command

**Background**
As noted in the previous section, an incident commander has a wide span of discretionary authority for making risk management decisions. There is no easy way to develop good judgment; some individuals are more capable in this regard than others. That factor should be considered in making promotions and assignments. Experience is always valuable, but gaining experience takes time, and experience can never address every situation an incident commander might face. Also, there is always a first time for any occurrence to happen.

Incident commanders can improve their abilities and self-confidence in making judgmental decisions through several different methods, beginning with a thorough understanding of the basic policies and their application to specific situations. Experience can be expanded and shared by:
- training
- studying reports of incidents that have occurred
- attending critiques
- carefully observing as many operations as possible.

Every fire officer and potential emergency incident commander must be prepared to make critical risk management decisions. An effective incident commander must be able to evaluate situations and predict outcomes, If the outcome of a situation is predictable, the incident commander can take action to minimize the exposure of operating personnel to danger. If the outcome is not predictable--or the indicators are not recognized, situations might be left entirely to chance.

To be prepared to make good risk management decisions, a command officer has to develop a foundation of knowledge, judgment and experience. The ability to predict outcomes and exercise good judgment must be based on either personal or shared experiences. The risk assessment process should be internalized well in advance of an incident in which its application might be a matter of life or death.

*Continued on next page*

**Experience: a job requirement**

One of the best ways to develop a base of experience is to consciously identify and evaluate the risk factors that apply to each emergency incident and to practice making judgmental risk management decisions. The conscious application of the risk evaluation process is a good start, but few individuals have the opportunity to personally experience a full range of situations before they are called upon to use their risk management skills.

A fire officer must be able to evaluate situations and predict their potential outcomes, particularly high risk situations that might result in very bad outcomes. If the danger is recognized, the incident commander should be able to determine the appropriate course of action to reduce, minimize, or completely avoid the risk. Judgment is important to differentiate risks that are acceptable from situations that are too dangerous.

**The need for judgment**

The application of risk management policies relies heavily on the judgment of officers and supervisors, as well as individual firefighters and emergency responders. A capable company officer or command officer must be able to predict what is likely to happen in a wide variety of situations and to weigh the risks against the potential benefits of different actions. Actual and potential dangers must be recognized, evaluated, and placed in perspective in relation to the three guidelines that define acceptable risk (referred to earlier). The ability of trained and experienced officers to make appropriate risk assessments and apply risk management policies requires good judgment.

Judgment is often associated with experience. Personal experience is probably the easiest way to develop good judgment, but in today's reality, a total reliance on experience as the basis for judgment is inadequate. There are far too many lessons to be learned about managing operational risk and too few occasions to develop the necessary experience. In addition, there is no assurance that the experience will come before the situation that requires the best judgment. There isn't time to wait for the experience to develop the judgment.

*Continued on next page*

| | |
|---|---|
| **Developing judgment** | A responsible officer must make a full spectrum of efforts to learn about the subject matter that forms the foundation for good judgment, including:<br>• taking courses that include operational safety<br>• reading books<br>• studying reports of incidents that have occurred in other jurisdictions<br>• attending critiques<br>• looking at buildings under construction and demolition<br>• looking at buildings where fires have occurred<br>• visiting properties to develop pre-fire plans<br>• simulation-based training/critiques.<br><br>Training for officers and incident commanders should provide a thorough understanding of the principles of operational risk management and the opportunity to practice their application in exercises and simulations. That type of training is essential to prepare individuals who might have to make critical decisions in situations they have never personally experienced.<br><br>Even those officers who respond to a large number of incidents generally face only a few situations that are truly challenging. The relatively large percentage of situations that are uncomplicated and easily controlled tends to make it even more difficult to rely on personal experience to recognize crises and act appropriately. As a result, many incident commanders rarely get enough "real world" experience to develop solid judgment and decision making skills. That's why training that provides the opportunity to develop command skills is critically important, Simulations--followed by thorough, detailed critiques provide essential learning experiences for all potential incident commanders.<br><br>Incident commanders must develop a discipline for managing incidents systematically, by consistently applying the process and the principles to all types of situations. The regular application of the process will help to develop the habit of approaching every situation in a standard manner. This discipline is essential to avoid the situation in which an incident commander, facing a critical incident, has to apply an unfamiliar process to fulfill his or her risk management responsibilities. |

*Continued on next page*

**Judgment under stress**
There are times when critical risk management decisions must be made under conditions of extreme stress, with incomplete information and with only seconds to evaluate the alternatives. An incident commander must be prepared to decide if a given situation justifies the exposure of responders to the degree of risk that would be involved in implementing different strategies. Those determinations must be based on judgment that allows an incident commander to apply the principles of operational risk management.

*Continued on next page*

**Predictability**

The ability to anticipate the outcome of emergencies is an essential component of a risk manager's effectiveness. Once a dangerous situation is recognized, it is usually possible to avoid or prevent the undesirable outcome.
As Gordon Graham, Risk Manager of the California Highway Patrol, has said, "If it is predictable, it is preventable."

An incident commander must have the ability to:
• recognize dangerous situations
• predict what could happen
• take action to avoid the predictable danger.

An incident commander should never find himself or herself in the position of having to say "I realized what could have happened, but all I could do was hope for the best." In many cases, the explanation for an unfortunate outcome has been "we recognized the danger, but we took a calculated risk." Unfortunately, in many situations, that means somebody recognized danger and yet did nothing to prevent the occurrence.

The incident commander should be pessimistic. When an incident commander recognizes a risk factor that could result in a bad outcome, he or she should make the avoidance of that bad outcome an integral part of the strategic plan. The incident commander should evaluate the steps that will either prevent it from happening or protect firefighters and emergency workers from harm if it does happen. Possible risk reduction or avoidance actions must be:
• considered in relation to the probability and potential consequences of the negative outcome, and
• balanced against the need to take action to control the emergency.

**Do not rely on good luck**

An incident commander should never rely on good luck to make an operation safe and to keep a predictable outcome from happening. Bad luck can ruin a good operation. There are some unavoidable risks that are virtually outside of anyone's control, Those are the inherent risks of conducting emergency operations. Fortune and misfortune definitely play a part in determining outcomes, but they should have an impact only in circumstances that are beyond the control of an incident commander.

*Continued on next page*

**Recognition of dangerous situations**

An incident commander must have the ability to quickly recognize and react to dangerous situations. The process of looking out for indications of danger should be a routine part of the incident management process. It extends to everyone involved in operations at the scene. The incident commander--and everyone else involved--should always be on the lookout for indications of potential danger.

**Balancing probabilities and consequences**

Risk management always involves balancing probabilities and consequences. There is at least a remote possibility that any burning building could collapse because of some undetected flaw in the construction. That said, it is also true that firefighters cannot effectively control structure fires if they stay out of every burning building. They must be able to recognize buildings that have characteristics that make them more susceptible to structural collapse. There is no excuse for having firefighters buried in the rubble of a building that exhibited recognizable warning signs or characteristics prior to collapse.

In emergency operations, it is often necessary to work with the recognition that something bad could happen. It would be impossible to take effective action without facing some danger. There are some overwhelming potential dangers that are always at least remote possibilities. A trained and experienced officer must develop the ability to recognize dangerous situations and predict outcomes. Once the danger is recognized, the risk assessment has to consider the probabilities and potential consequences to decide if the predictable outcome is unacceptable.

*Continued on next page*

**Training and education**

It is difficult to train an individual to exercise good judgment. Judgment is highly dependent on experience, which must be acquired and internalized, however, the fundamental principles and application of risk management can be learned. Training should also allow an individual to learn about different situations that have occurred, so that the indicators can be recognized--even if the individual has never had the personal experience.

Example:

A fire officer must know that lightweight wood truss construction floors and roofs are likely to collapse without warning if the trusses become involved in a fire--even if he or she has never actually seen it happen. An officer must know about the risk, recognize the type of construction, and know where it is likely to be encountered. Similarly, an officer must know about the risk of a BLEVE, even if he or she has never encountered a burning propane tank. Fire officers should also learn to beware of buildings with entrances on multiple levels that might contuse companies operating on different floor levels. They should recognize occupancies that are likely to have hazardous contents.

Important experience can be acquired through training that gives an officer the opportunity to relate risk factors and indicators to predictable outcomes. Proven methods include:
- classes
- lectures
- books
- videotapes and other media
- simulations and critiques

Standard texts such as Francis L. Brannigan's Building Construction for the Fire Service provide essential information that enables officers to recognize high risk situations.

Fire officers should also study published incident reports--particularly reports about incidents that result in injuries, fatalities or close calls--to determine significant factors in each case. Doing that should prepare an officer to quickly recognize situations that indicate an increased level of risk, such as buildings with lightweight construction components or components that are susceptible to sudden failure.

*Continued on next page*

# Preparation for Incident Command, Continued

**Critiques**

Participation in incident critiques should be an important part of the learning process for everyone who responds to emergency incidents. Fire officers should carefully review every incident in which they have been involved and use each new experience to expand their personal risk evaluation skills. Looking back--after having seen the outcome--at each significant incident allows the participants to focus on the accuracy of their observations and their analysis of the situation.

The learning experience in a critique can be shared by others who were not involved in the incident or in the direction of the operation. One of the most important values of a critique comes from the capacity to replay--in light of the known outcome--the thought process that went into making strategic and tactical decisions. Officers who were involved in the decision making process should discuss what they saw and how they interpreted the situation, then explain the decisions that were made based on that information. A critique provides an opportunity for others to share the experience of using information and observations to predict outcomes.

**Observation**

Personal experience should also be supplemented by observing as many incidents as possible. It is often revealing to stand back at a safe distance and watch what is happening at an emergency scene. The "big picture" might be quite different from the narrow perspective that is available to the participants, who might be seeing only a small part of the situation. Monitoring radio traffic also helps to compare visual observations with information reported over the radio from different positions and vantage points. A fire officer should also develop the skill to look for risk indicators in occupancies and locations where they may have to respond at some time in the future. Pre-fire plans should be developed for high risk occupancies, particularly where the risk indicators are not easily recognized. An observant risk evaluator should always be looking for indications of risk factors.

*Continued on next page*

**Observation** (continued)

Some of the most important characteristics of fire behavior can often be interpreted by observing the behavior of smoke and flames and their relationship to the building. Visual clues may also provide important evidence of structural weaknesses. The incident commander should try to establish the command post at a location that provides a good view of a fire scene. If the incident commander cannot make a personal 360 degree size-up, he or she should assign personnel to make a full visual survey and report back to the command post.

**Practice**

The incident commander must have a well developed sense of the priorities that will have to be applied to actual situations, based on actual incidents that have occurred and on hypothetical situations that can be imagined or predicted. Aspiring incident commanders should try to apply this value system to situations that are likely to occur in the particular area where they may be called upon to make those decisions. Working out situations in advance can prove to be extremely valuable when the real situations have to be confronted.

**Investigations**

The investigation of accidents that occur during emergency operations is an essential component of operational risk management, Those events:
- illustrate the weaknesses of established approaches
- identify areas in which improvements are needed.

Although some injuries are unavoidable, they should never be considered acceptable. All injuries should be investigated. The results should be used to reduce the probability of a reoccurrence. In fact, the primary purpose of any safety investigation should always be to reduce the probability of future accidents.

*Continued on next page*

**Investigations** (continued)

Line of duty deaths must always be thoroughly investigated to determine the cause and the steps that are necessary to absolutely keep it from happening again, **A** fire department should be prepared to immediately launch a full scale safety investigation when a death or injury occurs. (The *Guide for the Investigation of a Line of Duty Death,* published by IAFC and the *Standard Protocol for an Autopsy for a Firefighter Line of Duty Death,* published by USFA, are recommended references.)

*Continued on next page*

**An example of an investigative report**

The following section includes portions of a Major Fires Investigation Report developed by the U.S. Fire Administration after a fire in Seattle claimed the lives of four firefighters on January 5, 1995. The Report demonstrates the complexity of a major incident and the difficulties that can be encountered in trying to identify, evaluate and manage the risks that might be present.

## RISK ASSESSMENT

The importance of conducting fire suppression operations with a "risk management approach" is emphasized in NFPA *Standard 7500,* the *Standard for a Fire Department Occupational Safety and Health Program.* The application of the risk management concept to this incident is particularly significant. The risk assessment must be based on a combination of factors including a size-up of the structure and the fire conditions. The Incident Commander seldom has complete knowledge of all the potential risk factors during the early stages of an incident. Operations must often be initiated based on the best information that is available and then adjusted as additional information is obtained.

The fire occurred in a building that was occupied by two operating businesses and was reported by a tenant who detected smoke. The situation observed on arrival appeared to be an exterior fire extending to the interior. The structure appeared to have many of the characteristics of a heavy timber structure. These observations would support the initiation of an interior attack. The initial strategic plan was an offensive attack, intended to keep the exterior fire from penetrating into the building.

An east to west attack direction was identified by the first arriving officer and approved by the Acting Deputy Chief who assumed command of the incident. Additional companies were assigned to the west side to prevent extension of the fire and specifically directed to avoid a conflict with the interior attack crews. The plan was effectively communicated to all of the operating crews. At this stage of the operation, it was not known or suspected that that the fire was actually in the basement or that the basement contained a structural element that would result in a sudden floor collapse.

A pre-fire plan might have made the Incident Commander and other officers more aware of the arrangement of the structure and could have resulted in earlier recognition of the fire in the basement. It is questionable whether the "flaw" in the structure that resulted in a sudden floor collapse would have been recognized when a pre-fire plan was developed.

*Continued on next page*

**An example of an investigative report** (continued)

As the incident progressed, the attack plan appeared to be successful. The visible exterior fire was controlled and there were no reports of significant interior involvement from the attack crews. The fact that they had not encountered any significant fire involvement on the ground floor was not reported, so it was assumed that the attack plan was working.

The main body of fire had been located in the basement, under the attack teams, by a company assigned to the west side of the fire, but this information was not reported back to the Incident Commander, because its significance was not recognized. This critical factor could have caused the Incident Commander to evacuate the interior crews before the floor collapsed under their feet.

**Progress Reports** - The Incident Commander often has to depend on progress reports and information from other observers to evaluate the effectiveness of the attack plan. One of the primary responsibilities of company officers and division or sector officers is to keep the Incident Commander informed through regular progress reports. The Incident Commander should be informed immediately of any factors that could impact on the overall strategy for the incident, To evaluate interior progress, Division Supervisors must either go inside, where the companies are operating, or depend on company officers to keep them informed with accurate reports.

At this incident the assigned Division Supervisors did not transmit any progress reports, positive or negative, to the Incident Commander. The company officers who were leading the interior attack teams did not provide any progress reports or information on interior conditions to their Division Supervisor, who was located outside the building. He could see that the interior was heavily charged with smoke, but did not know that the companies were encountering very little fire inside. He did not have any information to provide progress reports to the Incident Commander and was not asked for a report.

The Division Supervisor on the west side saw the large fire in the basement, moments before the collapse occurred, however, the significance of this observation was not recognized and it was not reported. It was assumed that the Incident Commander knew about the fire in the basement The presence of interior attack crews immediately above was not known to the crews who found the fire.

In the absence of progress reports, the Incident Commander had to rely on other indicators to evaluate the effectiveness of the attack plan. The exterior fire was knocked down and there were no reports to suggest that the interior crews were having difficulties controlling the fire. These observations suggested that the strategic plan was working. A progress report from either Division Supervisor could have caused the other to recognize the inconsistency and would have alerted the Incident Commander to the problem.

*Continued on next page*

**An example of an investigative report** (continued)

**Battalion Chiefs** - The lack of progress reports appears to be related to the lack of aides or assistants to support the Division Supervisors. The Battalion Chiefs, who are normally assigned as Division Supervisors, do not have aides.

The Battalion Chiefs were overloaded, trying to direct operations and perform accountability functions at the same time. Seattle uses a "passport" accountability system, which requires a control point to be established outside the building, near each entry and exit point. The Battalion Chiefs were performing this function, as well as trying to monitor two separate radio channels. These responsibilities kept them outside, at secondary command posts, where they had to rely on company officers to inform them of interior conditions.

A Battalion Chief would need a "partner" to be able to go inside the building and at least one assistant to stay outside to perform the accountability function. This would require at least two assistants to be assigned to each Battalion Chief.

**Reconnaissance** - It was difficult to visually size-up the situation from a single vantage point. The configuration of the buildings made it difficult to interpret the interior arrangement without a pre-fire plan or a 360 degree size-up. Differences in grade levels made it difficult to lap all the way around the structure to determine the arrangement and access points.

The visible smoke and fire conditions were. difficult to relate to the actual location and magnitude of the fire inside the structure. All except one of the companies and command officers responding on the first alarm approached the fire from the same direction and all saw the large volume of flames against the west wall of the structure. This observation was interpreted as an exterior fire threatening to extend into the building.

The view from the Command Post was obstructed by trees and vegetation which made it difficult to visually size-up the structure. The area where the fire was venting from the basement to the exterior was below the street level and out of sight from the Command Post.

**Operating Companies** - The interior attack crews never encountered any significant interior fire involvement on the upper level; they found only a few spot fires near the floor which they quickly controlled. They did not realize the fire was in the basement, directly below them, because the concrete floor prevented smoke and flames from penetrating through.

The initial conditions were heavy smoke and moderate heat, which was consistent with their expectations. When rooftop ventilation was accomplished and the heated gases were released, the interior atmosphere cooled, which is usually an indicator of good progress. They believed that their efforts were successfully keeping the fire out of the building.

*Continued on next page*

**An example of an investigative report** (continued)

Engine 5 could not advance the line out onto the roof, because of the electrical power line that had dropped in their path. This kept them from closely examining the west wall of the two story section, which would have allowed them to see that the visible fire was actually venting out of the basement.

Engine 2 located the door into the fire area and determined that a large area of the basement was heavily involved in fire. Because of the assignment they had been given, they expected the attack crews to be pushing the fire toward them, so they prepared to defend their position. They did not recognize that no one else was aware of the large interior fire or that the interior crews were directly over it.

In retrospect, it can be determined that very different interpretations of the structure and the fire were being made from different vantage points. The individuals making these observations did not recognize the significance of their information to the Incident Commander. This emphasizes the value of a complete 360 degree size-up of the fire scene, as early as possible, by the Incident Commander or by an individual who can report in person to the Incident Commander with a "full picture" of the scene.

# Chapter Summary

**Key points**

In this Chapter, we have discussed how to apply risk management principles and practices to the planning, conduct, and evaluation of emergency response operations.

Clearly, the nature of emergency response itself carries with it some degree of inherent and unavoidable risk. Many of those risks can be readily identified and evaluated. However, no incident should be considered routine. That is why the presence of inherent risk only intensifies the need to implement sound risk management strategies at every level of an emergency response organization and before, during, and after every phase of emergency incident operations.

Experience and reality-based training help an emergency response officer develop the judgment incident command demands. Written policies and standards provide risk management guidelines. Personal protective equipment provides measurable margins of safety All that is true. Still, it takes diligent, watchful, ongoing application of consistent safety-minded practices--during every emergency response operation--to effectively reduce risk to acceptable levels and afford all personnel the highest levels of protection possible.

# Managing Information

# Chapter Overview

**Purpose**
The ability to assemble and process incident-related information is an essential skill for people who direct emergency operations. This Chapter presents several approaches to managing information to support operational risk management.

**In this Chapter** This Chapter includes the following topics:

# The Importance of Managing Information

**Assessing risk**  Risk assessment decisions have to be based on information about the risks that are present and significant in each situation, so that the incident commander can predict what is likely to happen. An individual who has accurate, timely, reliable and complete information is in a much better position to evaluate risk than one who is working with fragments of information and large quantities of confusion. The incident commander seldom knows everything about a situation at the beginning and sometimes does not find out about important factors until the operation is over.

**Two categories of information**  Information considerations for risk management fall into two interrelated categories:
- pre-fire (or pre-incident) planning and preparation
- information management and application during emergencies

Pre-fire (or pre-incident) planning is a **crucial** component of the risk management process Pre-fire planning activities are intended to gather, store and retrieve information that will be valuable when a fire or other emergency incident occurs.

What makes it so important? The two most important reasons to develop pre-incident plans are to:
- recognize hazards
- compile pertinent information about those hazards that will help an incident commander implement the risk management process.

The ability to acquire and manage information during an emergency incident is an important component of incident management and operational risk management. The incident commander's ability to achieve a safe and desirable outcome necessarily involves managing the risks that apply to each situation, In order to successfully manage risk and many other aspects of the incident, the incident commander needs to quickly and efficiently gather and process information.

# Pre-Incident Planning and Preparation

**Introduction**    Pre-fire planning and hazard awareness activities can greatly improve the ability of fire departments and other agencies to deal with operational risk. The more that is known about a property before an incident occurs, the easier it becomes to manage the incident and the related risk. An effective program will allow responders--in advance of an incident--to:
- recognize  risks
- fully  evaluate  them
- develop  appropriate  tactical  plans.

**The goal of pre-incident planning**    Pre-fire plans are often developed to support effective tactical operations, but their ability to provide hazard awareness information might be even more significant. That is because, first and foremost, a pre-fire plan should inform an incident commander of safety hazards. To do that, the information in the plan has to be:
- timely
- accurate
- accessible

Let's look just a little closer at that third point, accessibility. Emergency responders must be able to find the information they need when they need it. What good is accurate information--made available on a timely basis--if it is not also in a format readily **accessible** to those who need it the most?

**Three steps in pre-incident planning**    A pre-incident plan should make significant information available to the incident commander when the incident occurs. The most critical information relates to hazards that would be difficult or impossible to identify or evaluate at the time of the incident. Prior awareness may be the only way to account for these risks. Creating prior awareness takes three distinct steps:
- gathering  information
- recording and storing information
- retrieving  and  applying  information

In addition, the information should be reviewed at regular intervals and updated as necessary when changes occur.

Let's  discuss  each  step.

*Continued  on  next  page*

**Information gathering methods**

The most common method of developing pre-fire plan information is to assign the first-due company to gather information by touring target locations and other structures whose features present potential problems (such as lightweight truss construction). That approach serves the dual purpose of providing an on-site familiarization tour along with the gathering of information.

Other methods include:

- assigning a special team or group within the fire department to develop pre-fire plans
- entering information directly from information submitted for building, occupancy
- reviewing hazardous materials and special use permits
- requiring building owners to submit plans in a specific format that meets the requirements of the department. (Some cities have adopted ordinances that require owners of high rise buildings and hazardous materials occupancies to obtain fire department approval of detailed pre-fire plan guide books-- developed by licensed contractors in accordance with strict criteria.)

Whatever the method used, a standard format--such as the one that follows-- should be established to enter basic information about each location and particular facts that apply to certain categories of buildings or occupancies.

*Continued on next page*

# Pre-Incident Planning and Preparation, Continued

## Example of a Pre-Fire Plan Form

### General Information

Name of Occupancy_____Address_____

Response: Initial:_____ Working _____

Access Considerations:_____

Hydrant Locations: N_____ S_____ E _____ W _____

Key Box Location: _____Utility Shut-off Location _____

Fire Command Center:_____

Contact: Position_____ Phone_____

### Specific Information: Scope

### Size of Occupancy:

Height:_____Stories:_____Length: _____Total Area:_____

Occupancy

Classification: _____ Rating: _____

Fire Flow Required:             25% _____        50%_____        100% _____

### Construction:

Materials used: (1-4)

Type: Rating:_____ Frame: _____Support: _____

Fire Spread Concerns:_____

### Occupancy Risk:

Specific Concerns: _____

_____

### Protection (Built-in)

Detection/Alarm: _____ Panel Location: _____

Sprinkler Protection Pump: _____ Control Room: _____

Stand Pipe Coverage: _____

Fire Doors: _____

Ventilation Systems: _____

Rated Assemblies: _____

Other Considerations _____

_____

# Pre-Incident Planning and Preparation, Continued

## Example of a Pre-Fire Plan Form (Continued)

**Exposures:**

Life: _____

Exterior: _____

Interior: _____

Special Hazards: _____

Environmental: _____

Weather Factors: _____

Other Information: _____


Date of Pre-Fire Plan: _____By: _____

Date of Occupancy Sketch (See reverse side): _____ By: _____

*Continued on next page*

**The objective of site visits**

The objective of every pre-incident site visit should be to gather and organize as much information as possible about the locations where incidents might occur, particularly properties known to be complicated or hazardous.

Any information--for example, *NFPA 704, Standard on Fire Hazard Materials*--that would make an incident commander aware of a hazardous situation, should be considered important That is particularly true if the information would not be readily apparent or easily identified during an incident

**The value of site visits**

Fire suppression companies should regularly visit occupancies in their response areas to identify hazards and plan effective operations. Why? A pre-fire planning visit makes it possible--before an emergency occurs--to <u>detect, evaluate, and record</u> all types of hazards found at that site.

<u>Examples:</u>

The presence of construction components that are susceptible to sudden failure, such as lightweight wood truss roof and floor assemblies, should be prominently noted.

Unusual occupancy characteristics, such as hazardous materials storage or open shafts should be identified to alert responding companies and inform the incident commander of the hazards that are likely to be present.

In addition, familiarization visits allow a potential incident commander to update the recorded information and consider specific risks and the potential benefits of different strategies that might be employed if an incident were to occur at the location.

In the same way, officers should review pre-fire plan information from time to time, as a training exercise, to become familiar with the documented information and to ensure that it is routinely updated to reflect any changes that have occurred.

*Continued on next page*

**Storage, processing, and retrieval of information: system options**

Over the years, fire departments have developed many different systems with which they gather and store pre-fire plan information. In many cases, the information processing system has been structured around the capabilities of the hardware and software that happens to be available, rather than designing a system to meet specific needs. Many departments still rely on traditional paper-based systems. In recent years, though, information technology has made remarkable advances and significantly improved departments' gathering , storage and retrieval capabilities.

**Paper-based systems**

Some systems are based entirely on paper maps, diagrams and printed information. Frequently, information is kept in three-ring binders and carried in the cabs of fire apparatus. Other departments maintain filing cabinets in fire stations or communication centers. Some systems place pertinent information on the premises where an incident might occur. There it can be accessed by units when they arrive at the scene of a call. Storing a pre-fire plan on the premises is particularly useful for occupancies such as high rise buildings that have complex alarm and fire suppression systems

Obviously, the effectiveness of a paper-based system is limited by the ability to organize, store and retrieve information on paper.

*Continued on next page*

**Moving away from paper-based systems**

A second type of system uses more efficient information storage, such as microfiche readers that can store photographic images of tens of thousands of pages of information in a small space. Transferring information from paper to micro-fiche is unavoidably labor-intensive, and retrieval systems and equipment are relatively delicate. Despite those potential drawbacks, micro-fiche systems are extremely valuable when used in support of large scale operations in a high density urban environment, With a micro-fiche reader, an incident commander in a mobile command post vehicle can access enormous amounts of vital information.

Examples:

Block-by-block aerial photographs of a city

Hydrant and sewer maps of each block and adjacent area

Footprint and floor-plan drawings of individual buildings.

Typewritten supplementary information about:
• buildings
• occupancies
• specific processes
• special hazards

*Continued on next page*

127

**Applications of advanced information technology**

Advancements in technology have made it easier to store and process vast amounts of pre-fire planning information as electronic images in computer systems. It is now possible to store tens of thousands of pages on electronic media such as CD-ROM disks. The same two practical limitations that have affected the use of earlier technologies such as micro-fiche have an impact on electronic storage systems as well. They are:
- the effort required to gather and organize the information
- the availability of the appropriate hardware to retrieve it.

Several systems have been developed as components of computer aided dispatch (CAD) systems to automatically retrieve information associated with a specific address when a team is dispatched to an incident at that location. Capabilities currently existing range from New York City's system--which can print out a few lines to alert responding companies of specific hazards associated with each address--to the system in Fairfax County, Virginia, which can store several pages of text information and line drawings in a CAD system accessible from mobile digital terminals.

The newest versions of mobile digital terminals have significantly increased capabilities to store graphic and text information and can be coupled to CD-ROM units that can store thousands of plans.

*Continued on next page*

**Factors that drive gathering, storage, and retrieval decisions**

The purpose of pre-fire planning is to put information in the hands of an incident commander. Choices among gathering, storage, and retrieval systems must always be made with that purpose in mind.

One of the most important considerations in entering and managing information should be the easy retrieval of the information when it is needed to support an incident commander's planning and risk assessment priorities. The system should:

- be readily available
- be easy to use
- be accurate
- clearly emphasize the most critical factors.

Examples:

If explosives are stored in a building, the basic, drawing should include a prominent notice of that hazard. The incident commander should not have to look in a text page under "contents" to make that determination.

If a warehouse is used to store iron pipe fittings, that "low-risk" factor can be noted in the secondary information pages.

If a building's construction incorporates an unprotected lightweight wood truss roof, that should be prominently noted in a manner that draws the user's attention.

**Effective use of graphics**

Graphics presentations should use standard symbols and color coding to convey important information efficiently. Information that is less time-sensitive might be accessed by referencing text pages. Drawings should provide enough information to support operational needs. They should be "user friendly." More detailed drawings and information, less likely to be critical during an incident can be documented as secondary reference information.

# Managing and Applying Information During Emergencies

**Introduction**     An effective pre-fire plan system must be capable of putting the significant information at an incident commander's disposal when an incident occurs. The dispatch system should inform responding units when there is a pre-fire plan available for an occupancy and the records management system should make the information immediately available.

**Early command**     The principles of incident management emphasize the need for one person to be in overall command of every incident, from beginning to end, although the identity of that individual might change during the incident.

If action takes place before command is established, an incident commander has to start the process from behind to try to catch up with the situation. To evaluate exposure to risks, an incident commander must know where companies are operating and what they are doing. An incident commander who has to start up the command process after all the first alarm units are committed is likely to spend the most critical decision-making time just trying to figure out who is where doing what.

A delay in establishing command adds risk to a situation because it interferes with the prompt identification, evaluation and processing of the risk factors. Risks that have not been identified cannot be evaluated or managed. When effective command is initiated from the outset of operations, an incident commander should have effective control over all the action taking place and should be in a better position to focus on evaluating risks. When the identity of the incident commander changes, an orderly transfer of command process ensures that continuity of information is maintained. For that reason, the officer assuming command responsibility for an incident will, in many cases, keep the previous incident commander involved at the command post for that exact purpose.

*Continued on next page*

**The management process structures the flow of information**

The incident management process is also intended to structure the flow of information--up, down and through the entire chain of command. Company officers and sector/group/division supervisors move information up the chain of command, just as they move orders and instructions down the chain of command. Effective communications are essential to make the system work.

Officers at different levels in the organization routinely make risk evaluation and risk management decisions within the scope of their responsibilities. Those decisions must be coordinated with the strategic plan that is directed by the incident commander. They must also be coordinated with other units, when the recognized risks may impact on their tactics.

Although an incident management structure is intended to facilitate risk management and information flow, it can have the opposite effect if the officers assigned to supervise different areas focus entirely on their own specific assignments or functions and consequently fail to fulfill their responsibilities for coordinating actions with other groups. Coordinating between activities is often as important as supervision over individual functions. Supervisors must pay attention to what is happening around them as well as what is happening under their direct supervision.

*Continued on next page*

**Consequences of failure**

The failure to communicate significant information up and down the incident management structure is one of the most common breakdowns that leads to firefighter injuries and fatalities. Post-incident analysis of fatal incidents often reveals that critical observations were made by individuals on the scene, but the information was not effectively communicated. In many cases, the individual who had the information did not recognize its significance or realize that it was not known by the incident commander. In other cases, the information would only have been significant if its inconsistency with information coming from a different vantage point had been realized. If neither observer reports their observations, the inconsistency is unlikely to be recognized.

Example:

In several incidents, groups of firefighters have died when floors gave way under them while they were searching for, or trying to extinguish, fires on one floor level when the main body of fire was actually on the level below them. In some cases, other crews were working on the lower floor level, aware of the fire, but unaware that crews were working directly above the fire. At some incidents, individual firefighters noted factors that indicated the fire was below them--such as a very hot floor or small fires breaking out at the floor level--but did not recognize the significance of their observations until after the incident. Deaths might have been avoided if information had been communicated in a manner that caused the incident commander to recognize the situation and remove the crews who were operating directly over the fire.

*Continued on next page*

**Command support staff**

Risk management responsibilities might be compromised if officers within the incident command structure do not have anyone to assist them with support functions, including information management.

An officer responsible for supervising an area or function must be aware of tactical activities taking place in that area as well as the risk factors present. An officer who is preoccupied with managing information and resources could lose track of the actual activities occurring within his or her areas of responsibility. That might become a problem for the incident commander as well as for supervisors at other levels of the incident management structure.

Information management is recognized as an important component of incident command.

An <u>incident commander</u> should have one or more assistants at the command post to assist in managing information and handling communications. In the "fully expanded" ICS and IMS systems specific responsibilities are defined for information management units *(notably RES-STAT for maintaining resource status information and SIT-STAT for maintaining information related to the situation)*.

A <u>sector officer</u> often requires one or more staff assistants to help keep track of information (including accountability tracking) and to ensure that effective communications are maintained with the incident command. A standard system for managing information at a command post or within a sector is an important component of an effective incident management system.

*Continued on next page*

**Providing support**

In many fire departments, the responsibility for supporting the incident commander and sector officers used to be assigned to chiefs' aides who acted as both chauffeurs and staff assistants for chief officers. Budget constraints (and the image of an officer being chauffeured) have caused many fire departments to eliminate this position. A variety of other approaches have been developed to provide an equivalent capability to support command officers at the scene of an incident.

One option is to dispatch additional command officers to working incidents to assemble a command support staff who can effectively manage incidents. Some departments routinely dispatch as many as four chief officers to working one-alarm fires to fill pre-designated staff positions, including safety officer. They may be supplemented by a special command post vehicle that responds with its own crew to perform command support and information management functions.

Another option is to train and dispatch pre-designated companies to perform this function at working incidents.

**Command support systems**

To control the information necessary to manage an incident, fire departments should use systematic approaches, including:
- standardized tactical reference worksheets
- status boards
- accountability tracking systems

A standard approach allows for incidents to be expanded or contracted, for command to be transferred, and for the command structure to be adjusted, without losing track of important information.

*Continued on next page*

---

**System design
considerations**

All forms and systems should be designed for upward integration so that information developed early can be easily incorporated into an expanding command support system if the incident grows.

Information management systems should be user-friendly and should use color and format to visually assist the user in organizing information. Forms and worksheets can be designed to provide reminders and guide the user through complex situations, with check boxes to indicate that functions have been addressed and blanks to fill in information. Where appropriate, forms should include spaces for drawn-in diagrams and sketches or for pre-printed plans, maps, data sheets or printouts from computer systems. The walls of command post vehicles should be set up to locate the same information in the same place every time with "free form" space left for unusual information that could be important for unusual incidents.

All components of an information management system should fit together. Sector officers should have clipboards with grease pencils or writing instruments that can be used outdoors, regardless of the weather. When passport accountability systems are used, individual name tags should fit onto company or team boards, the company boards should fit onto sector boards and there should be designated spaces on the walls of command post vehicles to mount the sector boards. If information is stored in computers, it should be formatted to print out in a format that is compatible with other forms and worksheets.

Forms and worksheets should provide a record of an incident that will be valuable for critiques and analysis. Even so, the first priority should be to assist command officers and company officers in managing incidents. The officer who concentrates on filling out forms perfectly might fall into the trap of failing to manage the incident or the risks.

---

# Special Information Management Considerations

**Hazardous materials**

Fire departments should require occupancies that process, store or use hazardous materials might be required to submit inventory information and material safety data sheets (MSDS). Fire departments should have systems that make such information immediately available to units responding to a reported incident at that location. Pre-incident information should include:

- where hazardous materials will be found
- how it is stored
- how much of it is stored or used
- how it should be handled.

The information should also include:

- appropriate emergency actions
- personal protective clothing and equipment requirements
- evacuation requirementsand distances
- any other information that would be significant in the event of an emergency.

One of the advantages of pre-incident planning is that it provides for all the time needed to fully research a situation before it occurs, Instead of attempting to obtain and interpret information about a hazardous material that is already involved in an active incident, the research can be conducted in cooperation with the party responsible for the premises and with access to every available expert and source of information, A fire department might not be able to anticipate every situation that can occur with hazardous materials, but it should be able to assemble enough information to be prepared for most incidents that occur in fixed facilities.

*Continued on next page*

**Technical rescue planning**  Advance planning and practice can increase the safety of all types of technical rescue operations. Although such incidents can present many complex and unpredictable challenges, some factors are predictable. Standard operating procedures should be developed for specific types of operations and their application should be incorporated in pre-incident plans for location where those incidents can be anticipated.

Particular attention should be directed toward locations where the potential need for technical rescue operations can be anticipated. Such sites include:
- major construction sites
- bridges and tunnels
- popular recreational climbing areas
- swift water boating areas

It is also possible to develop a standard operating procedure for a specific type of technical rescue incident that can be applied at any location. A standard operating procedure is a generic pre-incident plan, which is applicable or adaptable to a variety of situations.

**Confined spaces**  Occupancies where workers occasionally enter areas or equipment that meets the definition of a confined space are required to make provisions for rescue. In addition, they must make the location available to the designated rescue team for familiarization and training. Confined space entry certificates are required for entrants and will provide essential information about the space and the hazards, When a fire department is made aware of this type of hazard, the full procedure that would be used for entry into the confined space should be researched, planned, documented and practiced. The rescue plan should be regularly reviewed for training purposes by the companies that would be expected to respond to the location.

In January, 1993, the Occupational Safety and Health Administration (OSHA) issued its final rule on confined space safety--1910.146. The regulation affects fire departments that respond to confined space incidents.

*Continued on next page*

**Arson intelligence**

One of the most difficult problems for fire departments is how to deal with information related to arson threats. It would be difficult to justify withholding this information from the fire suppression forces, because advance notice could allow them to identify hazards and avoid unnecessary risk if an anticipated fire actually occurs. In most cases, the appropriate action would be to try to prevent the arsonist from starting the fire to avoid the risk of deaths or injuries and the resulting destruction. Sometimes the increased presence of fire department units conducting pre-fire planning sufficiently deters and discourages an arsonist.

In some special cases, the appropriate law enforcement tactic might be to establish surveillance over the property in an effort to capture the arsonist in the act. That is particularly true when the information is confidential and the case is part of a major conspiracy. When that happens, fire suppression units that would respond to the location should be advised of the threat and the surveillance. They should review all available pre-fire plan information. A determination would have to be made if their presence at the location to develop a pre-fire plan would jeopardize the investigation. That is a special form of risk management decision that would occur in advance of a fire.

# Chapter Summary

**Key points**    To be effective, risk management decisions must be based on accurate, timely, and complete information about hazards that are present and significant.

Fire departments should design and implement management information systems that enable:
- pre-incident planning
- on-site incident risk management

Paper-based systems can be effective, but modern information technology is rapidly developing more efficient and more effective ways, such as electronic storage and retrieval systems that provide incident commanders with information that helps them recognize, evaluate, and control risks during emergencies.

Communication up, down, and across the incident chain of command is essential to make maximum use of all available information, facilitate the risk management process, and prevent loss and injury.

Incident commanders must be aware at all times of special circumstances and conditions at each emergency location that could affect risk management decisions.

# Appendix

# Organizations and Sources Providing Risk Management Information

American Society of Safety Engineers
1800 E. Oakton Street
Des Plaines, IL 60018

Emergency Services Insurance Program
McNeil Building
17-29 Main Street
P.O. Box 5670
Cortland, NY 13045

Fire Department Safety Officers Association
P.O. Box 149
Ashland, MA 01721

International Association of Fire Chiefs
4025 Fair Ridge Drive
Fairfax, VA 22033

International Association of Fire Fighters
Occupational Safety and Health Department
1750 New York Avenue, N.W.
Washington, DC 20006

Justice Institute of B.C. Fire Academy
4198 West 4th Avenue
Vancouver, British Columbia, V6R 4K1, Canada

Learning Resource Center
National Emergency Training Center
16825 South Seton Avenue
Emmitsburg, MD 21727

National Fire Academy
16825 South Seton Avenue
Emmitsburg, MD 21727

*Continued on next page*

## Organizations and Sources Providing Risk Management Information, Continued

National Fire Protection Association
Batterymarch Park
P.O. Box 9101
Quincy, MA 02269

National Institute for Occupational Safety and Health
Appalachian Laboratory for Occupational Safety and Health
944 Chestnut Ridge Road
Morgantown, WV 36505

National Safety Council
444 N. Michigan Avenue
Chicago, IL 60611

Occupational Safety and Health Administration
Publication Information (202) 219-9631
Public Information (202) 219-8151
(or contact your local or regional office)

Public Risk Management Association
1117 N. 19th Street
Suite 900
Arlington, VA 22209

Risk and Insurance Management Society, Inc.
655 Third Avenue
New York, NY 10017-5637

United States Fire Administration
16825 South Seton Avenue
Emmitsburg, MD 21727

Volunteer Firemen's Insurance Services, Inc.
P.O. Box 2726
York, PA 17405

# Select Bibliography

Brannigan, Francis L., 1993. *Building Construction for the Fire Service, Third Edition.* Quincy, MA: National Fire Protection Association.

Brunacini, Chief Alan V., March/April 1992. *Fire Command,* pp. 30, 93. A Game Plan Reduces Legal Risk.

Esenberg, Robert W., August 1987. *Fire Chief Magazine,* pp. 44-46. Starting a Risk Management Program.

Hudson, Kathryn, October 1986. *The International Fire Chief.* Liability: The Emergency Manager's Dilemma.

International Association of Fire Chiefs (IAFC) Management & Liability Committee, 1990. *Fire Service Risk Management Implementation Guide.*

Jenaway, William F., 1987. *Fire Department Loss Control,* Ashland, MA: International Society of Fire Service Instructors.

Jenaway, William F., Chairman, IAFC Risk Management and Liability Committee, April 23, 1990. *Product Liability Impacts Upon The Fire Chief.*

Jenaway, William F., Chairman, IAFC Risk Management and Liability Committee, 1991. *The Fire Chief's Risk Management Survey Says . . . . .*

Kipp, Jonathan D. and Loflin, Murrey E., 1996. *Emergency Incident Risk Management.* New York, NY.

Morris, G.P., Brunacini, N., and Whaley, L., April 1994. Fireground Accountability: The Phoenix System. *Fire Engineering*

National Fire Academy, 1984. Student Manual. *Fire Risk Analysis: A System's Approach.*

*NFPA 1500, 1992, Standard on Fire Department Occupational Safety and Health Program.* Quincy, MA: National Fire Protection Association.

*NFPA 1521, 1992. Standard for Fire Department Safety Officer.* Quincy, MA: National Fire Protection Association.

*Continued on next page*

## Select Bibliography, Continued

*NFPA 1561, 1990. Standard on Fire Department Incident Management System.* Quincy, MA: National Fire Protection Association.

Phoenix Fire Department, 1992. Standard Operating Procedure M.P. 206.06. *Incident Critique Sector.* Phoenix, AZ.

Teele, B.W., ed., 1993. *NFPA 1500 Handbook.* Quincy, MA: National Fire Protection Association.

Virginia Beach Fire Department, 1993. Standard Operating Procedure 019. *Incident Command Policy.* Virginia Beach, VA.

Volunteer Firemen's Insurance Services, 1987. *Introduction to Basic Loss Control for the Emergency Services.* York, PA.

Wilder, Steven, April 1995. *Health & Safety for Fire and Emergency Services Personnel, Volume 6, Issue 4,.* pp. 1, 4, 5. Risk Management in the Fire Service.

www.ingramcontent.com/pod-product-compliance
Lightning Source LLC
Chambersburg PA
CBHW081453170526
45166CB00008B/2413